Sustainable Knowledge

DOI: 10.1057/9781137303028

Other Palgrave Pivot titles

Antonio V. Menéndez Alarcón: **French and US Approaches to Foreign Policy**

Stephen Turner: American Sociology: **From Pre-Disciplinary to Post-Normal**

Ekaterina Dorodnykh: **Stock Market Integration: An International Perspective**

Bill Lucarelli: **Endgame for the Euro: A Critical History**

Mercedes Bunz: **The Silent Revolution: How Digitalization Transforms Knowledge, Work, Journalism and Politics Without Making Too Much Noise**

Mark Bracher **Educating for Cosmopolitanism: Lessons from Cognitive Science and Literature**

Carroll P. Kakel, III: **The Holocaust as Colonial Genocide: Hitler's 'Indian Wars' in the 'Wild East'**

Laura Linker: **Lucretian Thought in Late Stuart England: Debates about the Nature of the Soul**

Nicholas Birns: **Barbarian Memory: The Legacy of Early Medieval History in Early Modern Literature**

Adam Graycar and Tim Prenzler: **Understanding and Preventing Corruption**

Michael J. Pisani: **Consumption, Informal Markets, and the Underground Economy: Hispanic Consumption in South** Texas

Joan Marques: **Courage in the Twenty-First Century**

Samuel Tobin: **Portable Play in Everyday Life: The Nintendo DS**

George P. Smith: **Palliative Care and End-of-Life Decisions**

Majia Holmer Nadesan: **Fukushima and the Privatization of Risk**

Ian I. Mitroff, Lindan B. Hill, and Can M. Alpaslan: **Rethinking the Education Mess: A Systems Approach to Education Reform**

G. Douglas Atkins: **T.S. Eliot, Lancelot Andrewes, and the Word: Intersections of Literature and Christianity**

Emmeline Taylor: **Surveillance Schools: Security, Discipline and Control in Contemporary Education**

Daniel J. Hill and Daniel Whistler: **The Right to Wear Religious Symbols**

Donald Kirk: **Okinawa and Jeju: Bases of Discontent**

Sara Hsu: **Lessons in Sustainable Development from China & Taiwan**

Paola Coletti: **Evidence for Public Policy Design: How to Learn from Best Practices**

Thomas Paul Bonfiglio: **Why Is English Literature? Language and Letters for the Twenty-First Century**

David D. Grafton, Joseph F. Duggan, and Jason Craige Harris (eds): **Christian-Muslim Relations in the Anglican and Lutheran Communions**

Anthony B. Pinn: **What Has the Black Church to Do with Public Life?**

Catherine Conybeare: **The Laughter of Sarah: Biblical Exegesis, Feminist Theory, and the Laughter of Delight**

Peter D. Blair: **Congress's Own Think Tank: Learning from the Legacy of the Office of Technology Assessment (1973–1995)**

Daniel Tröhler: **Pestalozzi and the Educationalization of the World**

DOI: 10.1057/9781137303028

palgrave▸pivot

Sustainable Knowledge: A Theory of Interdisciplinarity

Robert Frodeman

palgrave
macmillan

DOI: 10.1057/9781137303028

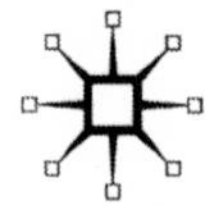

First published 2014 by
PALGRAVE MACMILLAN

Palgrave Macmillan in the UK is an imprint of Macmillan Publishers Limited, registered in England, company number 785998, of Houndmills, Basingstoke, Hampshire RG21 6XS.

Palgrave Macmillan in the US is a division of St Martin's Press LLC, 175 Fifth Avenue, New York, NY 10010.

Palgrave Macmillan is the global academic imprint of the above companies and has companies and representatives throughout the world.

Palgrave® and Macmillan® are registered trademarks in the United States, the United Kingdom, Europe and other countries.

ISBN: 978-1-137-30303-5 EPUB
ISBN: 978-1-137-30302-8 PDF
ISBN: 978-1-137-30301-1 Hardback

A catalogue record for this book is available from the British Library.

A catalog record for this book is available from the Library of Congress.

www.palgrave.com/pivot

DOI: 10.1057/9781137303028

Contents

DOI: 10.1057/9781137303028

Acknowledgements

The thoughts recorded here reflect a corporate effort, the result of years of thinking with Adam Briggle and James Britt Holbrook. While our work together extends back to 2004, since 2008 we have been finishing each other's thoughts at the Center for the Study of Interdisciplinarity (CSID). Of course, this particular version of our argument is mine alone. My colleagues should not be held responsible for any infelicities of thought or expression.

Keith Wayne Brown, manager of CSID, has brought his own distinctive Socratic spirit to all of our efforts. Kelli Barr, our graduate research assistant over the last few years, has also made substantial contributions to our projects. Steve Fuller, Julie Thompson Klein, Michael Hoffmann, Michael O'Rourke, Jan Schmidt, and Mark Bullock have each challenged my thinking on questions of philosophy and interdisciplinarity. And a special thanks to Axiothea, who has been a fellow traveler on this intellectual journey wherever she has been on the planet.

Finally, my sincere appreciation to Eunice Nicholson, who has helped vet the ideas expressed here over many glasses of wine.

DOI: 10.1057/9781137303028

1
Introduction

Frodeman, Robert. *Sustainable Knowledge: A Theory of Interdisciplinarity*. Basingstoke: Palgrave Macmillan, 2014.
DOI: 10.1057/9781137303028.

I A parable

In *Walden*, Thoreau tells the story of an Indian who goes door to door in Concord selling the baskets he's woven. He finds no buyers: while the baskets were beautiful, the man had not taken the trouble to make them worth anything to his neighbors. Academics have taken a similar approach to knowledge. They too have produced objects of great subtlety and beauty. But in too many cases they have not tried to make their research relevant to anyone beyond a disciplinary cohort. They have mostly followed Thoreau's path: "instead of studying how to make it worth men's while to buy my baskets, I studied rather how to avoid the necessity of selling them."

Thoreau sketched out the basics of a disciplinary approach to knowledge production. It's an approach that, despite its considerable merits, is breaking down today. Increasingly academics find their productions criticized and dismissed, their work habits called to account, and their funding cut. Society, it seems, believes it is getting a poor return on its investment in university research.

Sustainable Knowledge offers an account of this trajectory, the growing quandary in academic knowledge production that reveals itself in terms of the demise of disciplinarity. It is a predicament that manifests itself in a number of ways: as a crisis of knowledge overproduction; as a lack of larger relevance and social applicability; and as the forfeiture of authority, autonomy, and status among academics of all types.

Take the crisis of overproduction. Books and articles are seldom read with the care that's gone into the writing. No one can possibly keep up with the volume of material being produced. Our earlier way of dealing with disciplinary overabundance, through subdivision, Adam Smith applied to academic life, has been swept away by the epistemc deluge. Faced with a super abundance of knowledge in every subfield we divide and divide again, while the problems we face are increasingly integrative in nature. As a result disciplinarity – that is, knowledge production that limits its responsibilities within disciplinary boundaries – has become ineffectual, anachronistic, and defunct. It remains to be seen if interdisciplinarity can be any more successful.

This by design is a short book. One should expect no less for a volume concerned with the overproduction of knowledge. I have tried to make each sentence worth the reading, and to limit the narrative to matters where I have something distinctive to add. There is a rich and varied

DOI: 10.1057/9781137303028

literature on interdisciplinarity that the reader can consult; there is no need to repeat the points made in those volumes.

II On the use and abuse of knowledge

> We would serve history only so far as it serves life; but to value its study beyond a certain point mutilates and degrades life.
>
> Nietzsche

In scholarly parlance 'interdisciplinarity' refers to the integration of knowledge across the disciplines. This contrasts with the side-by-side juxtaposition of different types of knowledge, what is known as multi-disciplinarity, and the coordination of knowledge production with parties beyond the ivy walls of the academy, which goes by the name of transdisciplinarity. In what follows, however, I will usually follow the common practice and use interdisciplinarity as an umbrella concept that includes all those approaches that take us beyond a disciplinary approach to things. This is in keeping with the social fact that interdisciplinarity usually refers to a vague but omnipresent feeling that something decisive has changed about academic knowledge.

In its basic usage, then, interdisciplinarity is about many things. But on my view interdisciplinarity has a core meaning: it is about the most anti-modern of ideas, the notion of limit.

Now, disciplinarity is also about the concept of limit. Disciplines rely on boundaries to block off most of the world in order to pursue infinite knowledge within a limited domain, ideally with no outside interference. Interdisciplinarity breaks down those boundaries, but at the cost of limitations to understanding and expertise. Both disciplinarity and interdisciplinarity play across the registers of finitude and infinity in a Hegelian fashion. Interdisciplinarity can thus be described as a form of dilettantism – but that's judging the matter from the perspective of disciplinary knowledge, which does not seek as a matter of course to draw out the larger connections between things. And so while true, this criticism of interdisciplinarity makes as much (or as little) sense as saying that disciplines are isolated. Of course they are; that's what constitutes their epistemological bona fides.

Nevertheless, interdisciplinarity is about limit in a quite telling way: interdisciplinary knowledge production is limited by its need to be relevant to a specific problem or need at hand. Whereas disciplinarity

DOI: 10.1057/9781137303028

outlines an infinite research project, a theoretical digging without end, disciplinarity is in this sense ungoverned.

These pages try to make sense of a set of intuitions concerning the future of academic knowledge. What is the rightful place of knowledge in our lives?[1] Is it possible to have too much as well as too little of the stuff? Should knowledge production be governed by an Aristotelian mean? And how do we manage academic knowledge production under the vastly changed circumstances of 21st-century life?

In part I take inspiration from Nietzsche's *On the Use and Abuse of History*, which asks similar questions about historical knowledge. Of course, across his entire *oeuvre* Nietzsche asks about the purpose of truth and the psychological wellsprings of knowledge:

> What really is this "Will to Truth" in us? ... Why not rather untruth? And uncertainty? Even ignorance? (*Beyond Good and Evil*, 1886/2003)

While Nietzsche's concerns often centered on individual psychology, the focus here is more political in nature, probing the structural and institutional elements of the search for knowledge.

This work also shares Nietzsche's suspicion of the often-pious evocation of the nobility of the search for knowledge. 'Curiosity-driven' research is celebrated for its serendipitous results, but not all surprises are salutary. Curiosity was once considered a dangerous temptation; in modern culture it has become, in Einstein's words, something holy. There is something wonderful about "following the bent of one's genius no matter how crooked" (Thoreau). But curiosity should not function as a means of shielding academics from recognizing their public responsibilities.

To put one of my main points straight up, I believe that the period of infinite, encapsulated, largely autonomous and *laissez faire* knowledge production that characterized the last 150 years is coming to an end. There will be no total cessation, of course, nor should there be. But for a number of reasons – budgets, the dangers of technoscience, cultural disruption, and the irony of increased knowledge bringing us increased uncertainty and ignorance – the question of epistemic limit is likely to force itself on our attention. The pursuit of knowledge is no longer an unambiguous good – if it ever was.

It will become clearer that society can suffer from too much knowledge as well as too little, and that we should question the assumption that the answer to every one of our challenges is: more knowledge. The age now

DOI: 10.1057/9781137303028

passing away may come to be viewed as an era of epistemorrhea. To frame the point somewhat differently, the problem isn't only the absolute amount of knowledge, but also how knowledge is (or is not) balanced with other neglected qualities such as courtesy, solidarity, and quietude.

I am aware of the self-contradictions that this argument is liable to. It is hard to speak of silence, or to advocate the end of advocacy. Similarly, this is a work of knowledge that seeks to question the further production of knowledge. Some wags may suggest that I should honor my own insight and lapse into silence. To be clear: I remain a fan of knowledge. I enjoy producing and consuming it. In fact, I have never been quite sure what to do with life other than try to learn more about it.

Nonetheless, things are badly out of whack today. We have unrealistic if not millennialist assumptions about what we can expect to gain from additional knowledge (see Ray Kurzweil). As IBM argues in its 'Smarter Planet' campaign, it certainly is possible to use knowledge better. But this should not distract us from the fact that much of life is not about processing information. It's about learning how to be kinder, more open-minded, and fairer. Like William Buckley – not typically one of my heroes – albeit in a different context, I too seek to stand athwart history and yell "stop!" Or at least "slow down," introducing into our conversation the question of whether we have enough, or too much knowledge.

It's a point that Bill Joy raised in his article 'Why the Future Doesn't Need Us' (Joy 2000). That article should be required reading for all students entering college. It could serve as a useful antidote to the endless boosterism of knowledge culture, which has never seen a problem that cannot be fixed through the creation of additional knowledge.

III The book

Sustainable Knowledge develops two overall themes. First, it offers an account of the university and of contemporary knowledge production framed in terms of four core concepts, those of disciplinarity, interdisciplinarity, transdisciplinarity, and dedisciplinarity. Second, it reimagines the role of philosophy both within the academy and across society.

Concerning the first, the modern university system as it was created in the late 19th century and developed across the 20th century was built upon a set of assumptions rooted in the notion of disciplinarity. Today the social, economic, epistemological, and technological

DOI: 10.1057/9781137303028

conditions that supported those assumptions are breaking down. This is generally recognized, which is why interdisciplinarity has become such a buzzword.

But accounts of these terms too often slip off the main point, for example, into discussions of methodology, when what is needed is a historical and cultural critique of the changing role that knowledge is playing in culture. This is what I seek to offer here. I argue that knowledge production needs to be framed in environmental terms. Knowledge production has become unsustainable; academic sustainability forms an essential element in any serious attempt at sustainability overall. This is why the fourth chapter is titled Sustainability rather than Transdisciplinarity.

Second, I argue that the changes affecting the knowledge industry today highlight the inadequacy – indeed, the historical absurdity – of 20th-century philosophy and the humanities generally. We live in a deeply, irretrievably technological age; in what might be seen as an irony but which naturally follows, philosophy and the humanities have never been so necessary to our personal and public lives. But at this very moment the humanities have never been so marginalized. The humanities themselves are largely responsible for this situation: as they have constituted themselves across the 20th century, philosophy and the humanities have become peculiarly ill-suited to addressing the challenges we face both within the academy and abroad in society. Difficult as it may be to try to reorient the situation at this late date, there is nothing else to do but try. The humanities must reinvent themselves, taking on the Socratic task of providing a historical and philosophical critique of society.

These themes are elaborated upon across four chapters, respectively titled Disciplinarity, Interdisciplinarity, Sustainability, and Dedisciplinarity.

Disciplinarity offers an account of the disciplinary regime of knowledge production – its origins, conceptual assumptions, and current breakdown. What defines disciplinarity, both historically and conceptually? What purposes did it serve? What are the forces that are bringing about its end? This analysis emphasizes a point that is generally neglected, the crucial role played by the concept of peer review, which has functioned as the principle of governance of the disciplinary academy.

Interdisciplinarity provides an exposition of the concept of interdisciplinarity. I survey the strengths and weaknesses of some of the leading thinking on the concept over the past 30 years. The chapter also explores

DOI: 10.1057/9781137303028

two central motifs to interdisciplinarity as it has developed in the late 20th and into the 21st century, the concepts of method and rigor. In response, I offer an account of interdisciplinary virtues that is rooted in the thought of Aristotle and Heidegger.

Sustainability explores how we can best understand the concept of transdisciplinarity – actually a more crucial term than interdisciplinarity in understanding our current situation. I frame my argument in environmental terms, arguing that sustainability should become the master trope of transdisciplinarity. This section concludes with the claim that the central element of transdisciplinarity, aka the coproduction of knowledge, implies the recognition of limits to knowledge production – necessarily a repugnant notion for the academic status quo.

Dedisciplinarity argues for the dedisciplining of philosophy, and by extension the humanities. Disciplinary philosophy constitutes a category mistake: philosophy is not, or at least should not be exclusively a regional ontology (as are the sciences). I offer an account of the evolution of this mistake, and discuss the power and relevance that a dedisciplined approach to philosophy can have to culture at wide – with the added bonus of opening up new job possibilities for philosophers and humanists generally. Toward that end I propose the notions of the philosopher bureaucrat and the field philosopher as innovations necessary for refurbishing the role of the philosopher in society.

The book then concludes with an *Epilogue* that offers a narrative of the personal origins of the argument presented here.

This argument covers a great deal of ground. I hope that its brevity increases its readability, although I remember Kant's point that many a book would have been a briefer read if it had been longer. I have at points sacrificed depth of detail and scholarship for breadth of scope (what strikes me as a fair tradeoff). Part of the ideal of the pre- and post-disciplinary philosopher is to be provocative, to usefully and artfully outrage. This work will be a success if it spurs useful refutation as well as elaboration.

Note

1 I often hear of the need to distinguish between data, information, and knowledge. The only distinction that strikes me as pertinent is the one between any and all of these three and wisdom. The neglected question is: what is the relation between knowledge and living well?

DOI: 10.1057/9781137303028

Bibliography

Joy, B., 2000. "Why the Future Does Not Need Us." *Wired* Magazine 8.04, at http://www.wired.com/wired/archive/8.04/joy.html, accessed October 1, 2013.

Nietzsche, F., 1873/2010. *The Use and Abuse of History*. Cosimo Classics: New York.

Nietzsche, F., 1886/2003. *Beyond Good and Evil*, translated by R.J. Hollingdale. New York, NY, USA: Penguin Press.

DOI: 10.1057/9781137303028

2
Disciplinarity

Abstract: *This chapter offers an account of the disciplinary regime, its origins, conceptual assumptions, and current breakdown. What defines disciplinarity, both historically and conceptually? What purposes did it serve? What are the forces that are bringing it to an end? This analysis emphasizes a point that is generally neglected: the crucial role played by the concept of peer review, which has functioned as the principle of governance of the disciplinary academy.*

Frodeman, Robert. *Sustainable Knowledge: A Theory of Interdisciplinarity*. Basingstoke: Palgrave Macmillan, 2014.
DOI: 10.1057/9781137303028.

I Our situation

It had been a couple years since the last check up. So I make an appointment; have the blood work done; and show up for my visit.

Once there I am poked and prodded. We have the normal conversation about cholesterol, diet, exercise, and lifestyle. All is reasonably well. The session is wrapping up when my doctor says he has one thing more to talk to me about.

"Your PSA came back ... your number was 4.5."

I know that PSA (prostate-specific antigen) is a test of some kind for prostate cancer.

"So, what's that mean?"

"It means you need to make an appointment with an urologist."

Hmmm. But I do as I am told. Two weeks later I find myself at the urologist's office. He's a nice guy. I have trouble with doctors – they grow impatient with all the questions I ask. On a tight schedule and all that. But he takes his time with me.

He tells me that 25–30% of the men whose PSA is similar to mine will be found to have prostate cancer. You find out whether you have cancer through a biopsy of the prostate. If it turns out to be cancerous there are a variety of treatment options, up to and including the removal of the prostate.

This leaves me a little dismayed. I ask about the horror outcomes of prostate cancer – incontinence and impotency (death is out there as well, but somehow that seems a little less ... relevant). My urologist allows that all of these are possibilities. But with all the new treatment options, outcomes can be better. As for impotency, he volunteers, "don't worry. I can give you a hard-on that will be like a length of pipe." It is evidently just a matter of hydraulics. I believe I'm supposed to be comforted by this.

We schedule the biopsy for a couple weeks hence.

For some reason – perhaps the unpleasantness of the subject or the seeming logic of finding out whether or not I have cancer – I don't study up on the topic. Then, on the evening before the biopsy, I happen upon an article in the *New York Times*: "New Prostate Cancer Tests Could Reduce False Alarms" (NYTs 2). The article introduces me to a contentious debate about the usefulness of PSA tests. It turns out that there are questions about "whether healthy men should be screened for prostate cancer" at all.

DOI: 10.1057/9781137303028

This article links to an earlier *New York Times* piece from October 2011 (NYTs 1). This earlier piece focused on the announcement by a federal advisory board. Called the United States Preventive Services Task Force (USPSTF), the board had issued a Preliminary Recommendation that advised against taking the PSA test. Their meta-analysis of a number of studies examining results from treatment determined that "far more men were harmed by unnecessary biopsies and treatments than were saved from dying of cancer." Part of the issue is that in many cases prostate cancer is so slow growing that men die *with* prostate cancer, asymptomatically, rather than from prostate cancer.

The board also stated that the "benefit of preventing prostate cancer mortality in 1 man for every 1,000 men screened over a decade" should be weighed "against the known potential harms associated with screening and treatment" (i.e., discomfort, infection, incontinence, impotency, and death). Their summary statement is clear: "The USPSTF recommends against PSA-based screening for prostate cancer."

All this information is found in a few minutes. I dig around the Web a little more. I discover a response from the urologist's professional organ the American Urological Association (AUA). On May 21, 2012, in response to the USPSTF Final Recommendation against testing, the AUA had pushed back hard:

> The American Urological Association (AUA) is outraged at the USPSTF's failure to amend its recommendations on prostate cancer testing to more adequately reflect the benefits of the prostate-specific antigen (PSA) test in the diagnosis of prostate cancer. It is inappropriate and irresponsible to issue a blanket statement against PSA testing, particularly for at-risk populations, such as African American men. Men who are in good health and have more than a 10–15 year life expectancy should have the choice to be tested and not discouraged from doing so. There is strong evidence that PSA testing saves lives. The randomized trials used by the USPSTF do, in fact, show a benefit to patients. (AUA 2012)

The largest advocacy group for prostate cancer survivors in the US had responded in a similar vein. Thomas Kirk, the president of the prostate survivors group US TOO (of the blue rather than pink ribbons), is quoted in the *Times* article as saying: "The bottom line is that this is the best test we have, and the answer can't be, 'Don't get tested.'"

A little more time on the Web turns up opinions on both sides of the controversy. The debate is the sociological correlate of Newton's third law: for each expert there is an equal and opposite expert. I run across

DOI: 10.1057/9781137303028

a statement by a prominent urologist at Johns Hopkins University who claims it is foolish and dangerous to not take the PSA test – and then a quote, by another urologist, also at Johns Hopkins, who says precisely the opposite.

What to do? I waver, but then cancel my biopsy at 6 am the next morning.

That afternoon the telephone rings. It's my urologist. He is not happy with me. I am letting fear get in the way of a rational decision-making process. I protest: this is not a matter of emotion overcoming reason. I understand that my life may be at stake.

My urologist: "what possible harm is there in finding out if you have cancer?"

Me: "Well, setting aside the cost, unpleasantness, and the possibility of complications from the biopsy, from what I read, any subsequent treatment may be worse than the consequences of the disease."

The tension on the line is palpable. He is not pleased to be debating medical decision-making with me. And when I read him the USPSTF statement that "the Task Force concludes that many men are harmed as a result of prostate cancer screening and few, if any, benefit" (USPSTF 1) he explodes:

> "Those guys are a bunch of clowns. There's not a urologist in the entire group."

He's right about that: the Task Force is co-chaired by a pediatrician and a professor of family and community medicine. Other members include specialists in gerontology, public health, epidemiology, biostatistics, health management and policy, and a number of other areas.[1]

My urologist:

> "I have 25 years experience as an urologist; who are you going to believe?"

II Disciplinary expertise, transdisciplinary complications

It was clear that my urologist meant this as a rhetorical question. The issue of who to believe, or whose judgment to follow, shouldn't even come up. The problem was of my own making, caused by my indiscriminate reading around on the Internet. ("You call off the biopsy because of an article you find in the *New York Times*?") I lack the expertise to judge

DOI: 10.1057/9781137303028

these matters. Best is to leave it in the hands of the experts – in this case, urologists such as him. (Never mind that urologists were also disagreeing on these points.)

In fact, my urologist was implicitly making an entirely different point from what he intended. For the decision really *was* mine: which expert do I choose to believe? While the subject was science, or medicine, there was a type of fundamentalism at play here. The image of science is one of Newtonian certainty. The *reality* of scientific research is that it functions as a process of constant disputation. The Newtonian image is a useful fiction for scientists as well as politicians. Both are well served by the patina of certainty: the scientists retain their authority and all the benefits that flow from it, and politicians can refer to 'what science tells us' as a conversation stopper. Certainty, after all, brooks no debate or opposition.

My conversation with him also relied on the commonsense distinction between expertise and lay knowledge. In fact, society depends upon it: our lives are filled with expert opinions. (Try to buy a bicycle, a stereo, or a vacuum: every item has become a domain of expertise.) The AUA framed the debate over PSA tests in these terms, seeing the controversy as a matter where the expertise of scientific specialists is imperiled by political interference.

In 2012 the AUA joined with House Republicans to undercut the autonomy of the USPSTF. Describing the Task Force as an instance of governmental overreach, Representative Marsha Blackburn, Republican from Tennessee, sponsored House bill H.R. 5998. The bill would mandate a "greater role for specialists and advocacy groups in developing guidelines" while "eliminating the Department of Health and Human Services' secretarial discretion to withhold Medicare funding for interventions that lack convincing evidence for benefit."[2] Best to leave decisions in the hands of the experts.

As it happened, H.R. 5998 died in committee. But the controversy didn't. As the first *Times* article had noted:

> As the P.S.A. test has grown in popularity, the devastating consequences of the biopsies and treatments that often flow from the test have become increasingly apparent. From 1986 through 2005, one million men received surgery, radiation therapy or both who would not have been treated without a P.S.A. test, according to the task force. Among them, at least 5,000 died soon after surgery and 10,000 to 70,000 suffered serious complications. Half had persistent blood in their semen, and 200,000 to 300,000 suffered impotence, incontinence or both. As a result of these complications, Richard

DOI: 10.1057/9781137303028

> J. Ablin, who in 1970 discovered a prostate-specific antigen, has called its widespread use a "public health disaster". (NYTs 1)

Then, as I began to write up this account, I discovered that the AUA had changed its stance. From a May 3, 2013 *Times* article:

> In a major shift, the American Urological Association has pulled back its strong support of prostate cancer screening, saying that the testing should be considered primarily by men aged 55 to 69.
> The association had staunchly defended the benefits of screening men with the prostate test, even after a government advisory committee, the United States Preventive Services Task Force, said in 2011 that healthy men should not be screened because far more men would be harmed by unnecessary prostate cancer treatments than would be saved from death.
> But in new guidelines issued Friday, the urology association says that routine screening is no longer recommended for men 40 to 54 years old who are at average risk of getting prostate cancer. Screening is also not recommended for men 70 and older.
> The guidelines say men 55 to 69 should discuss the benefits and harms of screening with their doctors. And if they do choose screening, an interval of two years rather than annually would be better. (NYTs 3)

Quite a shift! (The new guidelines were issued on a Friday, the day traditionally reserved for disseminating news that one wants to bury.)

But what about the AUA's earlier position? Returning to the AUA website, I discover that the AUA memo opposing the Task Force's recommendation was now gone. In its place one finds only: "Sorry, the page you requested was not found."

But, of course, nothing is truly gone from the Internet. Checking the Internet archive *Wayback Machine*, I find the original memo quoted above, as well as the earlier AUA videos denouncing the USPSTF's recommendations. It seems that the experts had changed their minds.

In the meantime, men are still given PSA tests as part of their physical exams, without being asked or being apprised of these controversies.

III The end of disciplinarity

> Disciplinarity is now an impossibility, both ideologically and practicably.
>
> Paul Forman, 2012

I was tempted to tell my urologist about the book I was writing, for we had slipped onto questions concerning the current state of knowledge

DOI: 10.1057/9781137303028

production. He and his professional association the AUA were making a disciplinary claim: questions surrounding prostate cancer should be addressed by experts in that specialty. The Task Force, on the other hand, took a multidisciplinary or interdisciplinary perspective: the benefits of a given treatment could only be adequately evaluated by the intersecting insights of a number of different specialties.

The debate over PSA tests encapsulates the array of challenges we face today in making use of knowledge: identifying the proper role of science in decision-making; navigating the difficulties in properly parameterizing a problem; recognizing the limitations of expertise; dealing with the epistemic and social consequences of huge and often contradictory amounts of information instantly accessible via the Internet; and adapting to the progressive loss of authority and autonomy among knowledge professionals of all stripes. The medical profession is but one example of the problems facing the knowledge class today.

It would have been nice to pursue these points with my urologist. But I didn't try. In part it was his sarcasm in response to my raising questions about my treatment. But the problem runs much deeper than that. The medical system is predicated on the model of 'shut up and do as you are told'. This isn't a matter of personal temperament; there isn't time for anything more. Doctor visits are budgeted at eight minutes per patient, precluding the possibility of dialogue and the active involvement of the patient in his or her care. Moreover, questions of liability arise when responsibility for medical decision-making is distributed across multiple sources. No wonder his impatience at my 'woolgathering': it presupposed a world that could not exist.

But this does not change the fact that patients can and will Google their medical concerns. Epistemic worlds are colliding, with real world consequences. Medical knowledge is no longer the exclusive province of doctors, obtained through an arduous medical education. It is now used, and misused, by everyone with an Internet connection. It's a whole new epistemic world out there; we are going to have to adapt or be run over.

Forman is thus correct that disciplinarity has now become an impossibility. Many of us act as if we have not gotten the memo. Nevertheless, this is nothing less than a political revolution in knowledge. The regime of knowledge that has characterized the past 125 years is coming to a close; and given that knowledge is power, the consequences will be portentous. The control that universities once exercised over knowledge, in terms of production, dissemination, and certification, is ending. Knowledge is

DOI: 10.1057/9781137303028

now instantly available on handheld devices from an infinite number of sources, and most of them are non-academic in nature. It's a revolution driven by technological innovation that will drive the university and society at large toward major epistemic and cultural change.

Nonetheless disciplinarity will be difficult to shake, not least because it primarily exists in our minds. As Forman notes, one of the most striking aspects of the disciplinary era was its lack of curiosity about its own nature. This lack of curiosity may have been intentional: better to not ask questions if the answers are likely to lead to "the elimination of almost every gratifying feature of the professorial profession" (Forman 2012). Instead, in an assumption that did not match the subtlety of their own disciplinary research, academics have supposed that disciplines were rooted in the facts of the world. Then, through the natural evolution of research, academic standards would rise as epistemic differentiation inevitably occurred.

It would occur because the world broke into natural kinds. So, for instance, on the one side we find rocks, and on the other life. The academy mirrors these natural divisions by giving us geology and biology. (Limestone suggests what is problematic about that.[3]) Alternatively, academics claimed that disciplines were based in a distinctive set of methods, tools, or perspectives. Such accounts offered up smaller versions of the demarcation problem, the philosophical project of trying to find a clear line distinguishing science from non-science. After considerable effort philosophers of science abandoned the attempt to identify a set of characteristics that allow us to mark science off from non-science (cf. Laudan 1983). Disciplines have been no more successful at the same project.

A 'discipline' is typically assumed to be an epistemological category.[4] Indeed, the Ur-text of interdisciplinarity, the 1972 OECD *Interdisciplinarity, Problems of Teaching and Research in Universities*, delineates "seven criterion levels for defining given disciplines on epistemological grounds ... for instance, zoology is concerned with animals; botany with plants" (Heckhausen 1972, p. 83). Granted, there is a specific epistemic content to a discipline. And one could even say that academic knowledge consists of nothing other than epistemological activity, since the job of the academic, scientist or humanist, is to make arguments of one type or another. But this is a red herring. The mistake comes in thinking that since academics are epistemologists, then the grouping of

DOI: 10.1057/9781137303028

different academic specialties within a given discipline forms a coherent epistemic whole distinct from other disciplines.

With greater success, Stephen Turner has argued that the definition of a discipline is largely nominal: a given discipline is constituted by what we have decided to gather together *as* a discipline (Turner 2000). What else unifies the four aspects of anthropology – physical anthropology, cultural anthropology, archeology, and linguistics? Or makes aqueous geochemistry and vertebrate paleontology both part of geology? Or makes the history of philosophy part of philosophy rather than history?

Turner notes a second characteristic of disciplinarity: disciplines function as internal markets and intellectual cartels. Disciplines consist of departments that train people to work in departments at other universities that have the same (or a similar) name as the first department. This is not simply a tautology, and it is more than a point about hiring or even economic relations; most centrally it is a point about *audience*. The disciplinary university created an internal economy of knowledge in which attention to peers trumps relevance to external social developments, needs, or modes of thought. During the disciplinary age non-academics were set up as marginal players in the process of knowledge creation, certification, and dissemination.

There are myriad reasons for the breakdown of disciplinarity – technological, sociological, epistemic, which is not to say that disciplines themselves are in danger of going away. Disciplines will remain, even as disciplinarity loses its power; there are practical reasons for their existence, such as our need for people trained in particular skills. But in the age now dawning, engagement with non-academics will become central to university functions. That is, the university is losing its disciplinary autonomy. This is a reflection of the fact that the university is losing its status as the *locus classicus* of knowledge production.

Academics, once they awaken to this threat – they have not, quite – can be expected to react bitterly to the loss of their prized autonomy. (It will be interesting to see if they will embrace their generally leftist political orientation and organize, protesting threats to their intellectual property, as universities take possession of their lectures, distribute them as massive open online courses (MOOCs), and downsize faculty.) The crucial point, however, is that this shift highlights the unsustainability of the central fact of our current mode of knowledge production: infinity.

DOI: 10.1057/9781137303028

When knowledge production was primarily an intra-academic affair, it operated without an explicit governor (there were always implicit ones, such as limits of time and resources, although these were treated as mere inconveniences). For academics, every answer raised a new question, an effective infinity of research topics. But as knowledge production becomes explicitly tied to social needs for knowledge, infinite knowledge production becomes unsustainable. Or so I will argue. It is this, even more than the loss of autonomy, which threatens the status quo of the academy.

The point is at root an ecological one. The shift from a disciplinary to an interdisciplinary culture is a movement from infinity to limit. Our infinite, now unsustainable consumer desires have been buttressed by the infinite production of knowledge. The shift to interdisciplinary knowledge production should form part of a new worldview where we have learned to live within limits.

IV An abridged history of disciplinarity

Academics are trained in and spend their professional lives housed within disciplines. But their thoughts on the subject are remarkably undisciplined, which is to say unorganized and poorly thought through. What is a discipline? When were disciplines invented? Are all disciplines alike, at least in terms of their nature as a discipline? Or do areas of study vary in terms of the degree or type of disciplinarity their field admits of? Should every academic field of study be 'disciplined'? Or should some research proceed in a non-disciplinary manner? What would that even mean? Overall, what is the purpose or function of a discipline?

You would think that disciplinarians would be keenly interested in such questions. And that these points would be part of everyone's training, something that would be picked up on the way to a BA or a PhD. Instead, very few ask questions about the disciplinary matrix they find themselves within. Moreover, the literature on these questions is surprisingly scant. Even researchers on interdisciplinarity, well aware of the inadequacy of disciplinary knowledge, only touch upon these questions without making a full frontal attack.

Now, one *can* find a number of essays that offer a historical or philosophical account of disciplinarity.[5] And one can piece together the story from works on the history of science and of higher education.[6]

DOI: 10.1057/9781137303028

But there exists no substantial body of literature – dare I say a discipline? – that focuses on the intellectual history of disciplinarity. And as I just noted, of what there is, disciplinary academics have not taken up this work and integrated it into their own self-conception in any significant way.

Individual disciplines have existed in one sense or another from antiquity. Pick three at random: in geography, maps date from Babylon and geographic treatises from Hecataeus of Miletus (c. 5th century BCE). Beginning in the 15th century the Age of Discovery added new knowledge and showed the limitations of classical geography, paving the way for the creation of modern geography by Humboldt and Ritter. In chemistry, ancient civilizations knew how to extract metals from ores and to ferment beer and wine. Such practical knowledge was complemented by the theoretical approach of alchemy that sought to explain the nature and transformations of matter. Chemistry replaced alchemy after 1661, in the wake of Robert Boyle's *The Sceptical Chymist*. In economics, the topic was tied to politics for centuries (cf. Aristotle's *Politics*). In the 17th century, economics began to separate itself from political science with the development of political arithmatik, leading to the physiocrats and the founding of modern economics by Adam Smith.

So much for the history of individual disciplines, which is not my focus here. Concerning the concept of 'discipline', depending on one's definition the idea can also be traced back to the roots of Western culture. The concept is often taken as simply meaning the division of knowledge into units. One could see 'discipline' as being born in the transition from Plato to Aristotle. While his dialogues cover a range of topics, Plato named his dialogues in terms of people and social roles – the *Theaetetus* and the *Phaedo*, the *Sophist* and the *Statesman* – rather than in terms of epistemic categories. In contrast, Aristotle framed his treatises – that is, the works that have come down to us, his dialogues having been lost – in terms of regions of knowledge such as logic, physics, and poetry. Similarly, it is possible to view the medieval trivium (grammar, logic, and rhetoric) as a disciplinary division. The trivium (and the quadrivium of astronomy, music, arithmetic, and geometry) for centuries formed the basic taxonomy of a liberal arts education.

But understanding 'discipline' as merely a taxonomic or analytic category neglects historical, institutional, and political elements that are central to the concept. While disciplines have a history that stretches back to antiquity, discipline in the rich sense, that is, modern disciplinary

DOI: 10.1057/9781137303028

culture, is the creation of the late 19th century. Disciplinarity is a complex concept that includes:

- Management of knowledge through breaking materials into discrete domains;
- Development of a research rather than archival orientation;
- A premium placed on the new; and
- The creation of a credentialed and professionalized caste.

Most crucially, however, modern disciplinary culture institutionalized a particular rhetoric in terms of its choice of audience. All the elements just listed are important, but disciplinarity is centrally defined by its focus on disciplinary peers.

Note that, traditionally, university departments or 'faculties' did not consist of domains of specialists. From the founding of the first university (the University of Bologna, 1088, at least in name, if not in terms of something we would recognize today) through the 18th century, universities were commonly made up of four faculties – the three higher faculties of theology, medicine, law, and the fourth of philosophy or the humanities. Professors would cycle through these faculties across the course of their careers, moving from the lower faculty of philosophy to the three higher faculties (Clark, 2006). Such movement discouraged specialization and narrowly focused conversations and encouraged an orientation toward broad-based learning.

Similarly, while new discoveries and insights have always been made, across medieval and early modern times there was nothing resembling today's institutional research culture dedicated to producing new knowledge. Rather than the production and dissemination of new knowledge, professors lectured (from the Latin *lectus*, past participle of *legere*, to read). This was in part a matter of technology: with books at a premium the oral transmission of knowledge was a necessity. The power of the church played an important role as well, in that professors were expected to promote a set of perennial truths – a *philosophia perennis* – rooted in religious orthodoxy.

In fact, while certainly involving the universities, much of European intellectual life from the 14th through 18th centuries transpired elsewhere. Creative exchange manifested itself through the *Respublica literarum* or the Republic of Letters, a set of learned social interchanges based in the post and the salon. Grafton (2009) notes that when the phrase first appeared in the 15th century it functioned as a general description

DOI: 10.1057/9781137303028

of the literary world. But by the early 17th century it was used to evoke a European and even worldwide set of scholarly practices. In Winterer's description:

> This was the great age of the generalist, the polymath, the *polyhistor*, adept in multiple fields of learning, when one could still aspire to know everything. Learned people exchanged books, letters, journals, antiquities, and scientific objects; they organized learned societies, academies, universities, laboratories, botanical gardens, and cabinets. (Winterer 2012)

The Republic of Letters differed from university life in three ways. First, there was a democratic impulse to the Republic of Letters that was lacking in university life. Talk of a *republic* within which all met on an equal footing may have been aspirational rather than actual, but it provided a powerful contrast in a world where class and monarchy still ruled. Second, the social and transnational aspect of the Republic of Letters, extending even to America and the Far East and involving daily debates in salons and coffee houses, highlighted the fact that intellectual activity was abroad in the world rather than simply cloistered in the university. Third, the Republic of Letters was heterodox in a way that universities, tied to church and state, could not be. Iconoclasts such as Rousseau and Voltaire stood outside the official paths of power, and in fact regularly ran into trouble with the authorities. For example, Voltaire's *Letters on the English Nation* (1733) angered both the French church and government and led to his exile from Paris.

Communication in the Republic of Letters was primarily carried out through letters, hundreds of thousands of which survive to this day. And while such letters would certainly announce the latest scientific discoveries in, for example, optics or mathematics, this system of communication was not constituted along disciplinary lines. Conversation ranged widely across questions of science, politics, philosophy, and religion. Figures such as Newton and Galileo are anachronistically viewed as scientists, when their education and orientation was decidedly general in nature, rooted in the classics and constantly crossing into the literary, the social, and the theological, a fact characteristic of education until well into the 19th century.

The Enlightenment laid the groundwork for the establishment of the disciplines. "The famous Enlightenment projects in the division of knowledge often aimed, with greater or lesser explicitness, to provide thereby a new foundation for a unified scheme of knowledge" (McKeon 1994). Thus the *Encyclopedie*, comprising 28 volumes published across the years

DOI: 10.1057/9781137303028

from 1751 to 1772, consisted of 71,818 articles and 3,129 illustrations. The Encyclopedia prepared the way for disciplines by offering a taxonomy via an alphabetic ordering that distinguished discrete areas for research. Stichweh (2001) correctly notes that the Encyclopediasts were mainly taxonomists, ordering and archiving knowledge rather than producing new knowledge, but that this was an important step on the way to disciplinarity.

This alphabetic ordering – which of course is no ordering at all logically, and which in fact functioned as a way to reject a hierarchical ordering of knowledge – highlights another element necessary for the development of disciplinarity: the change in the epistemic status of religion. For disciplinarity to triumph required the retreat of theology. In the Enlightenment, religion went from being the queen of the sciences and the end of all knowledge to being one among many types of knowledge. Talk of lower and higher faculty, and knowledge, with philosophy and the arts constituting the lower, medicine and law ascending to theology as the highest, needed to pass from the scene before there could be disciplines. The depersonalization of knowledge was also necessary. Rather than knowledge being attached to the person and thus a matter of status or personal charisma, knowledge was increasingly grounded in data that the world gave up through empirical investigation. This in turn provided a criterion for distinguishing between different types of knowledge in terms of how empirical they were.

The ultimate cause of the change from the static, archival function of knowledge to a dynamic, differentiating, and productive function of knowledge lies in a shift in worldviews that marked the invention of the modern: the development of new technologies, the discovery of new lands, and a different understanding of time and human experience, now seen as progressive. New worlds and experiences that could not be contained within old categories. Similar factors are in play today, suggesting that we may be in an analogous period of massive change.

V Disciplinarity and peer review

To underline a central point here: disciplinarity is rooted in the choice of audience. Disciplinary scientists communicate with cohorts who focus on the same discrete subject areas, create journals that are reviewed by experts in the field, and review younger members of the cohort for

DOI: 10.1057/9781137303028

tenure and promotion. The general name for these processes is peer review. Disciplinarity is thus grounded in the notion of peer review.

Peer review offered an answer to the political question of academic governance. Disciplinary knowledge production has not had to explain or justify itself, except to disciplinary peers. It is an approach to knowledge that severs the connection between knowledge production and use: despite the regular issuance of promissory notes concerning the relevance of academic research, on a day-to-day basis researchers produce knowledge with little practical concern with the uses non-academics put that knowledge to. Otherwise, we should expect regular if not constant conversation with the users to take place.

The isolation of disciplinary knowledge from extra-academic influences has generally been presented as a virtue, for instance in iconic works such as Bush's *Science, the Endless Frontier* (1945) and Polanyi's "The Republic of Science" (1962). On such accounts, academics create a reservoir of knowledge that society can draw upon – but where inputs into the reservoir are kept separate from societal outcomes (Pielke and Byerly 1998). A significant amount of work in the policy literature across the past two decades has questioned the Bush-Polanyi paradigm – for example, Sarewitz 1996, Guston 2002, and Pielke 2012. But this has had little effect on the general understanding of the relation between science and politics. To take this work seriously would call into question the entire edifice of disciplinary understanding of knowledge production.[7]

Disciplinary knowledge managed the feat of being viewed as simultaneously relevant and apolitical. Knowledge was taken to be an infinite and inherently beneficial good, which was somehow automatically relevant to society at large. The positive effects of knowledge justified further knowledge production; negative effects were the results of political actors or unknown cultural forces. In either case, academics argued that they had a special justification for self-rule: their activities – the production of knowledge, and its dissemination via presentations, publications, and teaching – were so specialized and so important that ordinary people could not properly judge their work.

As a result, academics devised a means for evaluating themselves that answers one of the oldest questions of philosophy: "who guards the guardians?" The disciplinary regime answered the question just as Plato did:[8] the guardians get to govern themselves. Biologists are the only ones competent to judge work in biology, and only chemists evaluate the research of other chemists. Non-experts – whether within or outside the

DOI: 10.1057/9781137303028

academy – will only disrupt the process, leading to misguided or even disastrous results.[9] The crisis of the disciplinary academy today thus manifests itself in part as a crisis in peer review.

Disciplinarity is thus defined by an internal gaze. There are two criteria for disciplinary knowledge production: the ability to say something original, and proficiency at attracting, and passing muster with, one's disciplinary colleagues. The standards of evaluation for disciplines have also been internal. The result is a structure that exemplifies the principal-agent problem.

In the principal-agent problem, one party (the agent) acts on behalf of another, less knowledgeable party (the principal). Since the agent has more knowledge about the topic than the principal, the principal cannot fully monitor the actions or decisions of the agent. At the same time the agent has a natural tendency to look after her own interests rather than the interests of the principal. Translated into the life of the academy, disciplinary interests come to trump the priorities of society at large. This is even turned into an epistemic truth and corresponding virtue: so Stanley Fish can say concerning the humanities:

> To the question "of what use are the humanities?" the only honest answer is none whatsoever. And it is an answer that brings honor to its subject. Justification, after all, confers value on an activity from a perspective outside its performance. (Fish 2008)

Surely Fish means no *extra-disciplinary* use, rather than denying that humanistic works can offer private pleasure, personal meaning, even religious ecstasy. But Fish, like so many, dismisses the possibility of an extra-disciplinary utility to the humanities as 'careerist'.

Peers use their expert judgment to evaluate one another. This focus on expert judgment is now in the process of being undercut by the rise of metrics for academic work. Bibliometrics, even when 'disciplinary' in focus, are destructive to disciplinary self-governance. For it takes no expertise to judge one number as being higher than another. Metrics thus promote non-disciplinary, non-expert evaluation, by persons such as deans, provosts, state legislators, and the public. Of course, metrics themselves invariably rely on prior acts of peer-reviewed judgment. A metric may appear objective, but when we open the black box we find various types of political and epistemic judgment about what should be measured, in what way, and by whom. But this fact has not slowed down the drive toward metrics.

DOI: 10.1057/9781137303028

Peer review is an inherently fragile compact. Someone was bound to notice that experts often disagree about the facts of the case, and are susceptible to biases based upon the personal or professional advantages of one conclusion over another (cf. Chubin and Hackett 1990). Anyone who has had a manuscript or proposal reviewed is well aware of the arbitrariness of the reviewing process, where slight differences in academic background can determine the success or failure of a grant, where reviewers can always find a neglected perspective to highlight, or to almost willfully miss the main point of an argument. Such notorious problems set the stage for a move toward metrics once computing power gained sufficient strength.

But the move toward quantification is not the greatest recent threat to peer review. Of even greater import is the increasing focus on the broader societal impacts of scientific research. The demand for taking account of the 'broader impacts' of research threatens the entire edifice of peer review and thus of disciplinarity itself.

In the US, a 'broader impacts' criterion was first included within proposal review process at the National Science Foundation in 1997, when the term was paired with the disciplinary concept of 'intellectual merit'. Review panels at first ignored the new criterion, then declared it obscure or meaningless or alternatively a threat to the integrity of science, or finally that it represented nothing new at all (the latter is the reading of Rothenberg 2010). But over time, in response to pressure from Congress, the centrality of the broader impacts criterion has grown (Holbrook 2012). Similar pressures can be found abroad, for instance within the Research Councils of the United Kingdom and the European Commission. In the newest set of NSF guidelines, broader impacts are approaching parity with the criterion of intellectual merit (NSF 2012).

The rise of such societal criteria represents the dedisciplining of the peer-review process, as the choice of which scientific project to fund becomes more overtly political. (The point applies to basic research agencies both in the US and abroad; 'mission' agencies such as NASA have broader impacts concerns embedded within its originating legislation.) Basic science agencies worldwide have put a great deal of thought into the question of broader impacts. But none of them have confronted the radical nature of this addition, which portends the destruction of the very idea of peer review.

Note the predicament: biologists have a special claim to judging the biological aspects of research, but they have no particular expertise or

DOI: 10.1057/9781137303028

priority in judging the ethical, legal, or societal effects of that work. But if these last two criteria are tied together – that is, if the disciplinary separation of research from outcomes is repudiated, and intellectual merit is judged together with broader impact, then science loses its crucial social status as knowledge that lies beyond the realm of the political. And with that, the modernist project of identifying a space for truth outside values and politics falls to ruins.

Practically, there are a number of ways to deal with broader impacts elements on the level of a review (Holbrook and Frodeman 2011). Reviewers can be asked to first identify the intellectual merit of a proposal, and then separately evaluate its broader impacts. Or different reviewers can evaluate different aspects of the proposal: biologists reviewing the biological aspects, and sociologists or economists the social and economic aspects. The problem comes when broader impacts considerations function as anything more than as a tiebreaker in deliberations, a point now being reached at NSF. Then the intellectual incoherence of the entire edifice becomes manifest. For what could possibly count as the intellectual merit of an article or proposal except its impact? The difference between intellectual merit and broader impacts then comes down to which audience is being impacted – a disciplinary, or a transdisciplinary one.

VI Crisis in the disciplinary university

But the crisis within peer review is still largely an implicit one. To academics, the more obvious and pressing problem is the one about autonomy: disciplinary self-rule is ending as the academy falls more under the sway of society at large. This can be framed in terms of the rise of the neoliberal university, as pervasive currents of 20th-century political philosophy affect one of the last bastions of independence (Frodeman, Briggle, and Holbrook 2012). But the process is more simply the result of globalization and technoscientific advance.

As I've noted above, the situation for academics is analogous to the one faced by the medical profession. Patients today arrive at the office informed – or often, misinformed – about their maladies and ready to challenge the physician's authority and expertise. The physician's authority is undermined further by the fact that medicine is increasing about lifestyle choices rather than simple matters of health – that is, 'health' is increasingly becoming a matter of personal definition and opinion.

DOI: 10.1057/9781137303028

The result is a type of deprofessionalization, as an area of discrete expertise is invaded by exterior concerns. A similar process of deprofessionalization affects both MDs and PhDs: PhDs, like physicians, are losing their autonomy as larger cultural issues become inherent to what had been their discrete disciplinary domain. Put differently, the conditions of epistemic scarcity that had underlain their authority are dissolved by readily available data/information/knowledge.

The transition to a new era of knowledge production will be an ongoing process, and could play out in a variety of ways. On one account:

> In fifty years, if not much sooner, half of the roughly 4,500 colleges and universities now operating in the United States will have ceased to exist. The technology driving this change is already at work, and nothing can stop it. The future looks like this: Access to college-level education will be free for everyone; the residential college campus will become largely obsolete; tens of thousands of professors will lose their jobs; the bachelor's degree will become increasingly irrelevant; and ten years from now Harvard will enroll ten million students. (Harden 2013)

Of course, prognostications are often just wild guesses. It may be that 25% of colleges will disappear, or 90%. Or we could even see a boom in the creation of new universities, as has happened in China. Rather than Harvard or, Stanford, it may be an as – yet uncreated educational entity (such as the Minerva Project[10]) that becomes the high-end provider of Internet-based education. Instead of disappearing, the bachelor's degree might shrink to three years, or be converted to flexible time units rather than semesters.

Other accounts will emphasize other factors – the breakdown of peer review and the growth of metrics in the face of demands for greater social accountability; the changing ways to certify skills, such as the awarding of 'badges'; the unsustainability of mounting student debt; the defunding of the public sphere; or the dangers of untrammeled scientific and technological progress, which puts tremendous power in the hands of malicious or unstable individuals. But in outline, the point stands clear: the mode of academic knowledge production that characterized the period after 1880 is coming to an end. The university is becoming unbundled.

Codifying the state of things, the current crisis of the disciplinary academy can be summarized in terms of three points:

1 The cost of higher education continues to rise at rates higher than inflation, while Web-based technologies make possible a radical

DOI: 10.1057/9781137303028

reduction of the cost of education. This sets up conditions for what Christensen calls a "disruptive innovation" – a thorough restructuring of the knowledge system (Christensen and Eyring 2011). One sign of the future: in the fall of 2011 some 160,000 students worldwide took a computer science course online with a professor at Stanford. Similarly, in the spring of 2012 MIT and Harvard announced plans for creating an online platform to offer free courses from both universities. This platform will include science and engineering as well as humanities courses.

2 The rise of individualistic, neoliberal political philosophies across the Western world. In the aftermath of World War II, higher education was conceived as having larger social and political as well as economic purposes (Schrum 2007). Conservatives offered a non-economic defense for higher education: students go to college not only to get trained for a job, but also to become educated in democratic virtues as a bulwark against totalitarian ideologies. But with the fall of the Berlin Wall, governments no longer saw the political, cultural, and philosophic aspects of higher education as a public good. And if education was simply a personal economic good, the individuals who reaped the benefits of education should bear the costs.

3 Universities no longer control the creation and dissemination of knowledge; and one must wonder how much longer they will control the certification of knowledge. In the US, corporations now spend nearly triple the amount on research compared to public sources such as the US National Science Foundation (Science Progress 2011). IBM estimates that humans now generate 2.5 quintillion bytes of data *each day*, with 90% of the information ever created having been produced in the past two years (IBM 2013). The overabundance of information has created the new management challenge of 'big data'. Much of this falls within the domain of Web 2.0, that is, user-generated content, through billions of blogs, tweets, texts, pictures, and videos, as well as cell phone GPS signals and purchase-transaction records. The resulting infosphere has become a smog of information that one surfs, cherry-picks, or simply ignores.

The crisis of the disciplinary academy is thus rooted in the loss of control over knowledge. Of course, disciplines themselves are not ending. They will continue to serve a central function in higher education as a means

DOI: 10.1057/9781137303028

for presenting and managing knowledge. But the belief or assumption that disciplinary knowledge is sufficient, that it is the *end* of knowledge, is expiring. Academics face increased demands for accountability while their funding is being cut. At the same time there is a growing recognition that our problems are complex, exceeding disciplinary frames, and always involve an inseparable mix of facts and values. Academics have lost their monopoly over knowledge production; are losing their long-cherished autonomy; and are likely to soon lose their control over the process of knowledge certification as well.

There was an implicitly platonic dimension to disciplinarity. That is, time wasn't real within the academy, at least in terms of research. Since they were ostensibly disconnected from the exigencies of market and society, academics were at leisure to pursue their interests at their own pace. Of course, questions of time and tenure always played a role, but a non-temporal definition of 'rigor' – that is, epistemological precision and detail – ruled as the dominant criteria for what counted as academic excellence. The question of the possible relevance of this research to an outside audience was of little importance, and was often dismissed as 'dumbing down' or 'outreach'. So was the question of the timeliness in reporting out one's results. There was little extra-disciplinary pressure to produce 'just-in-time'. The economic costs of research – in terms of time and materials, and in terms of lost opportunities for relevance – remained a largely unacknowledged factor in the production of knowledge.

Much of this story is generally sensed, if not well grasped in terms of the details. And so the academy has come up with a slogan for the changes that need to occur to respond to these crises: interdisciplinarity.

Notes

1 The complete list can be found at http://www.uspreventiveservicestaskforce.org/members.htm.
2 http://www.healthbeatblog.com/2013/05/psa-testing-an-about-face/
3 Limestone in most cases consists of skeletal fragments of marine organisms such as coral, bivalves, or foraminifera.
4 There is also a characterological dimension to the term, denoting the disciplined soul, which was ascendant until the mid-20th century.
5 Some of the more notable include Stephen Turner, "What Are Disciplines? And How Is Interdisciplinarity Different?" (2000); Rudolf Stichweh, "History

DOI: 10.1057/9781137303028

of the Scientific Disciplines," (2001); Peter Weingart, "A Short History of Knowledge Formations" (2010); David R. Shumway and Ellen Messer-Davidow, "Disciplinarity: An Introduction" (1991); Robert Post, "Debating Disciplinarity" (2009); Richard McKeon, "The Origin of Interdisciplinary Studies" (1994); Paul Forman, "On the Historical Forms of Knowledge Production and Curation" (2012); and the body of work by Julie Thompson Klein (e.g., 1990, 1996, 2005).

6 See Veysey (1965) and Clark (2006). Paul Forman (2012) offers an account of disciplinarity that also notes the lack of scholarly self-consciousness on the subject.

7 One can also view such research and innovative institutional expressions, such as Arizona State, as the harbinger of significant change.

8 In so many words. The exact wording never appears in the *Republic*; the closest is Glaucon's comment in Book III that "it would be absurd that a guardian should need a guard." Quis custodiet ipsos custodes? is usually attributed to the Roman poet Juvenal from his *Satires*. Nonetheless the phrase sums up Socrates' concern in the *Republic* with the training of the souls of the guardians.

9 The question of who should count as a disciplinary peer in the case of philosophy is addressed by Frodeman, Holbrook, and Briggle (2012).

10 http://www.minervaproject.com/.

Bibliography

AUA, 2012. "AUA Speaks out against USPSTF Recommendations," at http://web.archive.org/web/20120928041539/ and http://www.auanet.org/content/homepage/homepage.cfm. Captured October 1, 2012, accessed September 13, 2013.

Bush, Vannevar, 1945. *Science, the Endless Frontier: A Report to the President on a Program for Postwar Scientific Research*. United States Government Printing Office, Washington.

Christensen, Clayton M., and Eyring, Henry, J., 2011. *The Innovative University: Changing the DNA of Higher Education from the Inside Out*. San Francisco, CA: Jossey-Bass Publishing.

Chubin, Daryl and Hackett, Edward, 1990. *Peerless Science: Peer Review and U.S. Science Policy*. State University of New York Press.

Clark, William, 2006. *Academic Charisma and the Origins of the Research University*. University of Chicago Press.

Fish, Stanley, 2008. "Will the Humanities Save Us?" *New York Times*, January 6.

DOI: 10.1057/9781137303028

Forman, Paul, 2012. "On the Historical Forms of Knowledge, Production and Curation," OSIRIS, 27: 56–97.

Frodeman, Robert, Briggle, Adam, and Holbrook, J. Britt, 2012. "Philosophy in the Age of Neoliberalism," *Social Epistemology*, vol. 25, no. 3–4: 311–330.

Grafton, Anthony, 2009. "A Sketch Map of a Lost Continent: The Republic of Letters," *Republics of Letters: A Journal for the Study of Knowledge, Politics, and the Arts*, vol. 1, no. 1, May 1.

Guston, David, 2002. *Between Politics and Science: Assuring the Integrity and Productivity of Research*. Cambridge University Press.

Harden, Nathan, 2013. "The End of the University as We Know It," *The American Interest*, vol. 8, no. 3, at http://www.the-american-interest.com/article.cfm?piece=1352

Heckhausen, Heinz, 1972. "Discipline and Interdisciplinarity," *Interdisciplinarity: Problems of Teaching and Research in Universities*. Organisation for Economic Co-operation and Development.

Holbrook, J. Britt, 2012. "Re-assessing the science – society relation: the case of the US National Science Foundation's broader impacts merit review criterion (1997–2011)," in Robert Frodeman, J. Britt Holbrook, Carl Mitcham, and Hong Xiaonan (eds), *Peer Review, Research Integrity, and the Governance of Science – Practice, Theory, and Current Discussions*. Beijing: People's Publishing House, pp. 328–362.

Holbrook, J. Britt and Frodeman, Robert, 2011. "Peer Review and the *Ex Ante* Assessment of Societal Impacts," *Research Evaluation*, vol. 20, no. 3, September: 239–246.

IBM 2013. Bringing Big Data to the Enterprise at http://www-01.ibm.com/software/data/bigdata/, accessed date February 16, 2013.

Joy, Bill, 2000. "Why the Future Doesn't Need Us," *Wired*, 8.04, April, 2000, at http://www.wired.com/wired/archive/8.04/joy.html.

Klein, Julie Thompson, 1990. *Interdisciplinarity: History, Theory, and Practice.* Detroit: Wayne State University Press.w

Klein, Julie Thompson, 1996. *Crossing Boundaries: Knowledge, Disciplinarities, and Interdisciplinarities in the Series on Knowledge: Disciplinarity and Beyond.* University Press of Virginia.

Klein, Julie Thompson, 2005. *Humanities, Culture, and Interdisciplinarity: The Changing American Academy.* State University of New York Press.

Laudan, Larry, 1983/1996. "The demise of the demarcation problem," in M. Ruse (ed.), *But Is It Science? The Philosophical Question in the*

DOI: 10.1057/9781137303028

Creation/Evolution Controversy. Amherst, NY: Prometheus Books, pp. 337–350.

Michael McKeon, 1994. "The Origins of Interdisciplinary Studies," *Eighteenth-Century Studies*, vol. 28, no. 1 Autumn: 17–28.

NSF, 2012. National Science Foundation, "Revised Merit Review Criteria Resources for the External Community," at http://www.nsf.gov/bfa/dias/policy/merit_review/resources.jsp.

NYYs 1, 2011. "U.S. Panel Says No to Prostate Screening for Healthy Men," by Gardiner Harris, *New York Times*, October 6.

NYTs 2, 2103. "New Prostate Cancer Tests Could Reduce False Alarms," by Andrew Pollack, *New York Times*, March 26.

NYTs 3, 2013. "Looser Guidelines Issued on Prostate Screening," by Andrew Pollack, *New York Times*, May 3.

O'Rourke, Michael and Crowley, Stephen J., 2012. "Philosophical Intervention and Cross-disciplinary Science: The Story of the Toolbox Project," *Synthese* [Published online 13, September, 2012], doi: 10.1007/s11229–012–0175-y. http://link.springer.com/article/10.1007%2Fs11229-012-0175-y, accessed on November 4, 2013

Pielke, R.A., Jr., 2012. "'Basic Research' as a Political Symbol," *Minerva*, vol. 50: 339–361.

Pielke, R.A., Jr. and Byerly, R. 1998. "Beyond Basic and Applied," *Physics Today*, vol. 51, no. 2: 42–46.

Polanyi, Michael, 1962. "The Republic of Science: Its Political and Economic Theory," *Minerva*, vol. 1: 54–74.

Rothenberg, Marc, 2010. "Making Judgments about Grant Proposals: A Brief History of the Merit Review Criteria at the National Science Foundation," *Technology and Innovation*, vol. 12: 189–195.

Sarewitz, Daniel, 1996. *Frontiers of Illusion*. Temple University Press.

Schrum, Ethan, 2007. "Establishing a Democratic Religion: Metaphysics and Democracy in the Debates over the President's Commission on Higher Education," *History of Education Quarterly*, vol. 47, no. 3 August.

Science Progress, 2011. *U.S. Scientific Research and Development 101*, at http://scienceprogress.org/2011/02/u-s-scientific-research-and-development-101/, accessed date February 16, 2013.

Shattuck, Roger, 1996. *Forbidden Knowledge: From Prometheus to Pornography*. New York City: St. Martin's Press.

Stichweh, Rudolph, 2001. "History of the Scientific Disciplines," *The International Encyclopedia of the Social and Behavioral Sciences*, vol. 20: 13727–13731.

DOI: 10.1057/9781137303028

Turner, Stephen, 2000. "What are disciplines? And how is interdisciplinarity different?" in Stehr, N. and Weingart, P. (eds), *Practising Interdisciplinarity*, University of Toronto Press, pp. 46–65.
Veysey, Laurence R., 1965. *The Emergence of the American University*. University of Chicago Press.
Weingart, Peter, 2010. "A Short History of Knowledge Formations," in Frodeman, R., Klein, J.T., and Mitcham, C. (eds), *The Oxford Handbook of Interdisciplinarity*, Oxford University Press, pp. 3–14.
Winterer, Caroline, 2012. "Where Is America in the Republic of Letters?" *Modern Intellectual History*, vol. 9, no 3, November: 597–623.

DOI: 10.1057/9781137303028

3
Interdisciplinarity

Abstract: *This chapter provides an exposition of the concept of interdisciplinarity. I explore the strengths and weaknesses of some of the leading thinking on the concept over the past 30 years. The chapter also explores two central motifs to interdisciplinarity as it has developed in the late 20th and into the 21st century, method and rigor. In response, I offer an account of interdisciplinary virtues that is rooted in the thought of Aristotle and Heidegger.*

Frodeman, Robert. *Sustainable Knowledge: A Theory of Interdisciplinarity*. Basingstoke: Palgrave Macmillan, 2014.
DOI: 10.1057/9781137303028.

DOI: 10.1057/9781137303028

> In the midst of this chopping sea of civilized life, such are the clouds and storms and quicksands and thousand-and-one items to be allowed for, that a man has to live, if he would not founder and go to the bottom and not make his port at all, by dead reckoning, and he must be a great calculator indeed who succeeds.
>
> Thoreau

I Defining interdisciplinarity

I argued in the previous chapter that disciplinary knowledge production is principally defined by the question of audience. The disciplinary impulse implies a turn inward, where the primary audience for and the set of auditors of knowledge production consist of a limited group of peers. These peers share insider information, perspectives, and interests that differ from outside groups. Disciplinarity is thus subject to a number of hazards, which can be largely described in terms of the principal-agent problem. Academics disregard the priorities of the audience that funds most academic research – the public – and are little inclined to take responsibility for the unknown and possibly dangerous downstream consequences of a mostly *laissez faire* process of knowledge production.

So what, then, of interdisciplinary approaches to knowledge production?

Interdisciplinarity is typically defined epistemologically, in terms of the 'blending' or 'integrating' of different types of disciplinary knowledge.[1] I have argued that the belief in disciplines as epistemic entities is chimerical. Correspondingly, I will define interdisciplinarity politically and rhetorically, in terms of audience. While present, epistemological concerns play a largely subsidiary role.

Take the commonplace distinction between multidisciplinarity and interdisciplinarity. The distinction between merely juxtaposing different bodies of knowledge, versus synthesizing them, is a real one. It's just that the distinction runs *through* disciplines just as much as between them. In geology, for instance, an account of the Rock Canyon Anticline in Colorado[2] will involve the integration (or not) of various types of data: structural, mineralogical, geochemical, and paleontological. We can also choose to integrate cultural, political, or ethical perspectives – or not. We can mix and match any or all of these perspectives. The paleontological can be combined with the cultural, for instance, in terms of the aesthetic qualities of fossils. What we choose to mix is a matter of interest and audience.

The simplest way to define interdisciplinarity is in terms of a focus outward, away from a group of peers. But this hardly captures the complexity of the situation. In fact, this turn outward has yet to be adequately conceptualized. The turn inward is relatively straightforward: one seeks to satisfy the demands of a community of peers that share many interests in common. But in turning outward, who exactly is one's audience? Even if we avoid its widest compass – 310 million people across the US, or the 7.1 billion people around the world – a non-peer audience will have any number of interests, none of which will be anywhere near as well ordered as those of a set of disciplinary peers. What would be the nature of such a (ever-changing) relationship, in terms of time, attention, and money? And how would that relationship affect one's ongoing relations with one's peer group, given that the latter remains in charge of matters such as hiring, tenure, and promotion? Overall, in an interdisciplinary context, how do we balance inside and outside?

II The dilemma

Mahatma Gandhi, when asked what he thought of Western civilization, replied "it would be a very good idea." There is a similarly untried aspect to the project we call interdisciplinarity. By 'untried' I mean that efforts to theorize interdisciplinarity have failed to judge interdisciplinarity on its own terms. Instead, interdisciplinarians have implicitly imported disciplinary standards and perspectives into their efforts.

It's no wonder; it's nearly impossible to avoid doing so. Take this book as an instance. Readers will quite reasonably come to it expecting a competent survey of the current state of thinking across the field of interdisciplinarity. But even in this limited intellectual world the volume of material is great enough that mastering and responding to the literature can absorb most of one's time and efforts. We are caught on the horns of a dilemma: the very act of responding to similar thinkers threatens to make the work into a forbidding 'insiders' account of limited interest to the larger world. Yet failure to take such thinkers into account ignores the community of researchers who care most about these topics.

Thus even as the interdisciplinary literature has sought to reach out to diverse areas, it has been subject to a stronger counterforce that drives

DOI: 10.1057/9781137303028

conversation inward. One needs to position oneself within the literature; one must take proper account of the arguments of other interdisciplinarians. Moreover, the dangers of not paying obeisance to the existing literature are political as well as epistemological. If not properly honored – arguments cited and built upon in ways that demonstrate one's allegiance to one's reference community – a writer risks paying the ultimate price: being ignored. This is not (or at least, need not be) a matter of pique; this is how disciplinary communities work. You must place your thinking within the ongoing debate of a reference community. The result is almost inevitably that one writes a monograph of interest to a small group of disciplinarians – in this case, of interdisciplinarity.

The community that studies interdisciplinarity has not been blind to this point. Weingart's "Interdisciplinarity, the Paradoxical Discourse" (2000) touches on some of these ironies. But where the attitude was once that we should fight against falling back to disciplinary ground, increasingly the response by interdisciplinarians has been to advocate the creation of a discipline of interdisciplinarity. The argument for disciplining interdisciplinarity is usually couched in terms of intellectual coherence and/or policy relevance, but the point is as much political and economic in nature. For it is difficult to gain stable university resources for occasional interdisciplinary efforts. Moreover, once gained, such resources as often are the first to disappear when the next round of budget cuts occur.

Thus even interdisciplinarians come to think and work within disciplinary communities. As Krohn notes,

> Whatever drives people into highly complex interdisciplinary projects – curiosity, social responsibility, or money – the need of manageable objects and presentable results in their reference community drives them out again. (Krohn 2010)

This need is present because of the institutional housing of academics. We need those manageable objects and presentable results because the whole system is set up to pin our careers on the judgment of disciplinary peers.

How do we navigate across this dilemma? How do we avoid capitulating to the disciplinary impulse that is so deeply woven into both our intellectual habits and our institutional structures? In my own case, in this chapter and book, I've responded by charting an idiosyncratic course across the contemporary landscape of interdisciplinarity. If this constitutes a 'method' it does so in the spirit of what Thoreau called dead reckoning.[3]

DOI: 10.1057/9781137303028

One should note that such an approach – whether in the case of the small community of interdisciplinarians, or within any other field of academia – is unlikely to find much intellectual support or institutional shelter. More likely it will be described as amateurish or dilettantish. We remain the disciplinary sons and daughters of Kant:

> All industries, crafts, and arts have gained by the division of labor ... Would not the whole of this learned industry be better off if those who are accustomed, as the public taste demands, to purvey a mixture of the empirical and the rational in all sorts of proportions ... were to be given warning about pursuing simultaneously two jobs which are quite different in their technique, and each of which perhaps requires a special talent that when combined with the other talent produces nothing but bungling? (1785)

Better off if we remained disciplinarians? Perhaps. But this ignores another possibility – that our background notions of method and rigor might themselves be incoherent, uncritically indebted to a disciplinary view of the world. After all, Kant's position contains a number of assumptions that have not worn very well. Ecology has taught us that knowledge does not consist of discrete packets of information, and the world is increasingly impatient with simply waiting for the learned industry to come to its own conclusions in its own time.

Consider this, then, a brief for the bunglers, or perhaps better said the dead reckoners, where the argument is taken to the disciplinarians.

III The two sides of interdisciplinarity[4]

Today one finds two predominant attitudes concerning interdisciplinarity, the booster and the skeptic. On the one side, interdisciplinarity is envisioned as the panacea for what ails both research and education. Within the academy calls for new interdisciplinary initiatives has become the conventional way to indicate one's reformist orientation. As a consequence, interdisciplinarity remains a hot topic and the literature continues to grow. Universities create interdisciplinary colleges and appoint senior administrative staff to manage an increasingly explicit interdisciplinary portfolio. As one review article noted, interdisciplinary approaches to research and education can be found in areas as diverse as poverty, public health, and leisure studies (Jacobs and Frickel 2009) – just about everywhere, in fact.

DOI: 10.1057/9781137303028

In fact, it is possible to see interdisciplinarity as having already taken over the university, at least nominally. Some claim that research today is essentially interdisciplinary in nature, with disciplinarity largely limited to the undergraduate experience of majors and textbooks and the like. More accurately, however, research expresses a kind of hyper-disciplinarity. That is, research is interdisciplinary only in a limited and topical sense: researchers, either individually or in groups, combine different backgrounds, perspectives, or data sets, but do so in terms of a particular *topos* or place, an approach that continues rather than challenges disciplinary mores (Frodeman and Mitcham 2007). While these often transient relationships only occasionally generate a new disciplinary field such as biochemistry, they are essentially nascent disciplines in terms of their goals and outlook.

At the same time, skepticism grows concerning the promise of interdisciplinarity. It is not obvious that interdisciplinary research does a better job than disciplinary research at addressing societal problems. Nor is it clear that interdisciplinary majors leave students better prepared for either the workplace or for life. (Disciplinary – that is, academic – outcomes are, of course, much easier to evaluate than inter- or transdisciplinary ones, since disciplines allow for comparisons between like and like. Part of the definition of a discipline is that there is an agreed upon means for evaluating work. The lack of such standards is what makes the question of how to evaluate the 'broader impacts' of research and education such a difficult one.) Calls for interdisciplinarity often signify sound and fury but little else. Individuals resist making the changes necessary for interdisciplinary success – or cannot decide what would constitute success – and institutions are not willing to commit significant resources to interdisciplinary projects. The term itself is thus in danger of becoming an empty signifier or shibboleth.

Today there is a growing set of researchers who, having made interdisciplinarity the focus of their research, are trying to take a more critical approach. These theoretical efforts are matched by a growing institutional dimension to the study of inter- and transdisciplinarity. Recent signs of this organizational focus include the 2008 founding of the first Center for the Study of Interdisciplinarity in the United States, at the University of North Texas; the 2010 creation of the Philosophy of/as Interdisciplinarity Network (PIN); and the 2011 founding of the International Network for Inter- and Transdisciplinarity (INIT). With a more undergraduate focus, the Association for Integrative Studies, founded in 1979, renamed itself

DOI: 10.1057/9781137303028

the Association for Interdisciplinary Studies in 2012 and continues to have annual meetings.

The first impulse of most of these researchers is to get rigorous about interdisciplinarity. Typically this leads to a discussion of method. Sensitive to complaints about dilettantism, their goal has been to identify a step-by-step procedure for integrating disciplinary knowledge and perspectives. They believe that this will increase the efficacy of academic work for addressing societal problems. Method, one of the purported signs of disciplinarity – purported, in that philosophers of science have failed to identify anything like a method to science other than organized common sense – is thus made central to interdisciplinarity.

But this is merely the most overt mimicking of disciplinarity. Over time, and with little sense of irony, interdisciplinarians have sought all the other accoutrements of a discipline: a canonical set of readings; insider conferences, journals, and associations; degree programs and even freestanding departments. For the vast majority of these researchers this is a normal, tacit, and even subconscious process which raises no conceptual issues. This is simply how academic research works. Few are as explicit about the process as Gabriele Bammer, who makes clear her goal of founding a new discipline of interdisciplinarity (Bammer 2013). But whether implicit or explicit, trends strongly point toward the creation of a new discipline of interdisciplinarity, with specialists whose expertise lies in the analysis of how to communicate with and integrate across the disciplines, and between the disciplines and society.

There is another model possible. It is where the interdisciplinarian goes feral, largely abandoning his or her disciplinary roots. It's an entrepreneurial approach where one circulates among a changing roster of partners – not too quickly, for experience and trust must be built up – with only occasional visits back to one's reference community to check in on new insights and to recruit fellow travelers. For instance, work on water issues in southwest Colorado for a few years, with geologists, hydrologists, policy-makers and stakeholders. When this project reaches its terminus, join a community group concerned with fracking within city limits. Such projects might last anywhere from six months to four years.

These possibilities are open to any academic. It is nonetheless striking how philosophers, who should be the native tribe of interdisciplinarians, and who so constantly invoke Socrates as their patron saint, so rarely embrace his choice of how to live the philosophic life: working in the agora, finding philosophical issues *in situ* and in *media res*.

DOI: 10.1057/9781137303028

Interdisciplinary discussions commonly proceed through the analysis of case studies. These case studies, which in fact constitute transdisciplinary rather than interdisciplinary research, are relied upon to keep one's thinking about interdisciplinarity honest and grounded in real-world experience. But here something peculiar happens: the case study becomes an example of a theoretical point rather than the main point of the research. The central point slips from addressing a particular problem out in the world toward becoming evidence in support of one or another theoretical point for the reference community of interdisciplinarians.

Both goals are legitimate, of course; I simply mean to highlight once again the gravitational pull of disciplinarity. For instance, in conversation once with a prominent researcher in interdisciplinarity where we were discussing how we could make our research more relevant to the larger world, I suggested bringing policy professionals into our meetings and deliberations. His response: let's not. We would have to "dumb down" the discussion for them.[5]

Evolution within the community of interdisciplinarians thus tracks the development of other fields that have sought to challenge disciplinary strictures. Take the field of science and technology studies (STS). Since its inception in the 1960s the field has steadily moved toward disciplinary status. This shift has itself been the subject of debate, sometimes framed in terms of 'low' and 'high church' STS (the latter identifying with the goal of becoming a discipline; cf. Jasanoff 2010). Efforts to stay low church, for example the STS program at Penn State, often end up being crippled by the lack of dedicated funding and tenure lines.

The field of applied ethics has followed a similar trajectory. Environmental ethics, born in the 1970s out of the desire to address pressing environmental problems, over time has become more and more recondite. Participation in discussions has become dependent on being well read in the literature. This perhaps explains why Eugene Hargrove, the founding editor of *Environmental Ethics*, titled an editorial 25 years after the journal's 1979 founding "What Went Wrong? Who's to Blame?" (Hargrove 2003) Following upon a 1998 editorial that sounded the same themes, Hargrove lamented the lack of effect that environmental ethics has had upon either environmental science or policy making. His explanation: environmental economics had occupied the space that environmental ethics might have. Hargrove did not raise the possibility that environmental ethics had become too insular and disciplined,

DOI: 10.1057/9781137303028

or whether *Environmental Ethics* had lived up to its self-description of being "an interdisciplinary journal."

Bioethics offers us something of an exception to the process of disciplinary capture. The field has been attentive to its institutional status and place in the wider world (e.g., Ackerman 1980). This may be partly attributable to the fact that bioethics has emerged from several disciplines (philosophy, medicine, law, nursing, political science, sociology, theology, and more), where environmental ethics was largely an offshoot of philosophy alone.

In his 1973 essay "Bioethics as a Discipline," Daniel Callahan emphasized that bioethics should be something useful to those who face real-world problems:

> the discipline of bioethics should be so designed, and its practitioners so trained, that it will directly – at whatever cost to disciplinary elegance – serve those physicians and biologists whose position demands that they make the practical decisions. (2007 [1973], p. 21)

Callahan saw bioethicists as having two choices when it came to addressing their difficulties in communicating with scientists and physicians. They could "stick to traditional notions of philosophical ... rigor" and continue "to mutter about the denseness and inanity" of their non-philosophical colleagues. Or they could adapt an extra-disciplinary definition of rigor:

> Not the adaptation of expediency or passivity in the face of careless thinking, but rather a perception that the kind of rigor required for bioethics may be of a different sort than that normally required for the traditional philosophical or scientific disciplines. (Callahan 2007 [1973], p. 19)

We will return to Callahan's point about rigor below.

Make no mistake: the story of disciplinary capture is not simply the story of personal failings. There is a logic at work here. Care in thinking requires serious study and a nuanced understanding of the literature. The development of a reference community is inevitable, and to a degree salutary. But by its nature it begins to exclude outsiders. The question is how and to what degree we can mitigate this process – and what this natural epistemic and political process says about the social function of knowledge and the relationship between expertise and democracy. On my reading, part of the neglected message of interdisciplinarity is that it points up such inherent limitations to knowledge in terms of its ability to address societal needs.

DOI: 10.1057/9781137303028

IV The question of method

Students of interdisciplinarity break into two camps. The dominant group focuses on questions of method, technique, and codification. A second cluster treats interdisciplinarity as cultural critique. In recent years,[6] the distinction between these two positions has become clearer, in part because of the work of a small group of philosophers who have explicitly focused on the philosophy of interdisciplinary (cf. Hoffman, Schmidt, and Nersessian, 2013).

The interdisciplinary methodists come in a number of varieties. The boldest claims are perhaps those made by William Newell, a leading member of the methodists. Newell sees scholarship since 1990 as having answered most of the central questions surrounding interdisciplinarity. His definition of interdisciplinary studies – "a process of answering a question, solving a problem, or addressing a topic that is too broad or complex to be dealt with adequately by a single discipline or profession" (Klein and Newell 1996) – assumes that disciplines are themselves coherent epistemic entities rooted in natural kinds.

In a 2007 essay Newell outlined a stepwise process for applying an interdisciplinary approach to complex problems:

NEWELL, 2007: THE STEPS IN THE INTERDISCIPLINARY RESEARCH PROCESS

A. Drawing on disciplinary perspectives

1. *Defining* the problem (question, topic, issue)
2. *Determining* the relevant disciplines (including interdisciplines and schools of thought)
3. *Developing* a working command of the relevant concepts, theories, and methods of each discipline
4. *Gathering* all relevant disciplinary knowledge
5. *Studying* the problem from the perspective of each discipline
6. *Generating* disciplinary insights into the problem

B. Integrating insights through the construction of a more comprehensive understanding

7. *Identifying* conflicts in insights by using disciplines to illuminate each other's assumptions, or by looking for different concepts with common meanings or concepts with different meanings, through which those insights are expressed

DOI: 10.1057/9781137303028

8. *Evaluating* assumptions and concepts in the context of a specific problem
9. *Resolving* conflicts by working toward a common vocabulary and set of assumptions
10. *Creating* common ground
11. *Identifying* (non-linear) linkages between variables studied by different disciplines
12. *Constructing* a new understanding of the problem
13. *Producing* a model (metaphor, theme) that captures the new understanding
14. *Testing* the understanding by attempting to solve the problem

Newell's list is stipulative: as with telling warring factions that they should 'come together and reach agreement on differences', it leaves the hard work undone. And so in the first item, there is no sense that problem definition is often an irredeemably political and axiological matter where there is no neutral answer to be had, and that additional theoretical resources may be necessary beyond telling people to 'define the problem'. Nor does Newell acknowledge the role that practical demands such as time or cost play in the design or functioning of interdisciplinary projects.[7]

Taking a similar tack, in 2013 Gabriele Bammer offers a book-length account describing the process of creating a discipline of interdisciplinarity – titled, aptly enough, *Disciplining Interdisciplinarity*. Her goal is laudable enough: "How can academic research enhance its contributions to addressing widespread poverty, global climate change, organized crime, escalating healthcare costs or the myriad other major problems facing human societies?" Her answer: through the development of a method for combining disciplines for problem-solving:

> there is no substantial, well-established, internationally accepted methodology. There are no standard procedures for deciding, for example, which disciplines to include, what each discipline will contribute or how the different findings will be melded together. (Bammer 2013)

Bammer suggests that a newly created field of Integration and Implementation Sciences (I2S) could function as a discipline on the model of statistics.

Bammer recognizes that this project calls for both theoretical and institutional elements. Theoretically, I2S consists of an intellectual architecture of three 'domains' combined with a five-question framework. The domains consist of

- Synthesizing disciplinary and stakeholder knowledge;
- Understanding and managing diverse unknowns; and
- Providing integrated research support for policy and practice change.

These domains are then framed by five questions:

1. What is the integrative applied research aiming to achieve and who is intended to benefit?
2. What is the integrative applied research dealing with – that is, which knowledge is synthesized, unknowns considered, and aspects of policy and practice targeted?
3. How is the integrative applied research undertaken (the knowledge synthesized, diverse unknowns understood and managed, and integrated research support provided), by whom and when?
4. What circumstances might influence the integrative applied research?
5. What is the result of the integrative applied research?

Like Newell, Bammer pursues the goal of identifying a uniform set of questions applicable to every interdisciplinary situation. But she matches her theoretical account with a call for a worldwide Manhattan Project or Human Genome Project-level effort to collect and collate information on thousands of interdisciplinary research projects, looking for "concepts, methods and case examples." One would expect positive results from such an endeavor – if Bammer and colleagues can get it off the ground. But the source of its value is unlikely to reside where she thinks. Rather than in the development of an interdisciplinary methodology, its value will more likely come from the sharing of a wealth of particular insights and rules of thumb that have developed in a piecemeal manner.

It's a point that Wolf Krohn (2010) has made. He begins with an epistemic question: to what degree does inter- and transdisciplinary work consist of a series of 'one-offs' that resist generalization? Might not the heterogeneity of inter- and transdisciplinary projects resist codification?

Krohn frames his discussion of interdisciplinarity in terms of two models of reasoning, one classically humanistic and the other scientific. Following Windelband, he calls these the idiographic and the nomothetic.

DOI: 10.1057/9781137303028

He describes the idiographic as humanistic in its urge to understand the individual case in its specificity, while the nomothetic is scientific in orientation, viewing the individual through the lens of the general and looking to extract lawlike abstractions. Of course these are ideal types. Geology, for instance, has a strong idiographic tendency, in that we are more likely to be trying to understand the specifics of the creation of the Grand Canyon rather than searching for general rules about canyon formation. Similarly, in the case of climate change, our central interest is in understanding what is happening to this planet now rather than the identification of lawlike behavior applicable to planets in general.

Krohn sees interdisciplinary processes as leaning decidedly toward the idiographic. In this he follows a point made by Gibbons *et al.* (1994), who spoke of the need to develop interdisciplinary competencies. Krohn does not deny that it is possible to become proficient at interdisciplinary activities, or to instruct others in how to do so. But he sees little point in the search for general principles:

> A historian who specializes in the founding of the United States of America usually does not wish to become a specialist for foundings in general, but builds his reputation on knowing everything about just this case and giving it an original and surprising interpretation. If he cared to analyse another founding – say of the Roman Empire, Brazil, or the European Union – neither factual knowledge nor interpretation schemata can be transferred from one to the other. (Krohn 2010)

Krohn may overstate the point. For example, experience in working with a variety of stakeholders within environmental restoration should give the interdisciplinarian a set of skills that would be particularly useful in other restoration projects, and could also be adapted to other contexts (e.g., urban studies). But his central insight holds.

Krohn's theme has been made by others in a variety of contexts. In *The Savage Mind* (1966) Levi-Strauss speaks of *bricolage*, where one reuses old materials in order to solve new problems, an experimental process of testing, trying, and playing around. And Foltz (2000) describes thinking as a type of reconnaissance, a semi-organized theoretical wandering that does not prescribe an interpretive structure for the world.

Foltz's account is influenced by Martin Heidegger, who goes largely unappreciated as a thinker of interdisciplinarity, even though much of his *oeuvre* functions as a critique of disciplinarity. On Heidegger's account, a methodology is the last thing you want in the search for truth – a point

DOI: 10.1057/9781137303028

that analytic philosophers of science have come around to, having largely given up on identifying a scientific method.

For Heidegger, the application of a methodology is a kind of theoretical brutality where we disregard the individuality of an event. The imposition of a method forces a given situation to live up to a pre-established standard rather than allowing the situation to suggest its own standard for evaluation. Methodism thus fundamentally misunderstands the nature of thinking, which is at root a kind of questioning. "Words are not terms, and thus are not like buckets and kegs from which we scoop a content that is not there. Words are wellsprings that must be found and dug up again and again, that easily cave in, but that at times also well up when least expected" (Heidegger 1968; quoted in Babich 2013).

Perhaps, in a given situation, the political stakes are particularly high. Perhaps a great deal is at stake in terms of cost or environmental protection. Perhaps a cultural legacy is at risk, or the matter is particularly religiously fraught. Newell's list – for example, "defining the problem", and "determining which disciplines have relevant information" – is not pointless; addressing a problem will require moments of introspection on topics such as those he lists. But someone who seeks to promote an interdisciplinary perspective on a problem needs to be able to do more than list propositions. Interdisciplinary collaborations call upon political, rhetorical, and psychological nuance.

While also methodologically oriented, O'Rourke and Crowley (2013) have created a workshop format that digs deeper into our subconscious. Called the Toolbox Project, they have run more than 100 workshops across the US where local team members spend several hours becoming more aware of the epistemic, metaphysical, and ethical assumptions embedded in their specific disciplinary approach to research. The workshop begins with the participants answering a 34-statement questionnaire. The questionnaire is divided into 'Epistemology' and 'Metaphysics' sections using a Likert scale ('values' questions are included in the Metaphysics section). Participants then engage in a free-form, 90 minute guided discussion of their answers, at the end of which they fill out the questionnaire for a second time.

A sample question from the Epistemology section:

14. Validation of evidence requires replication

Disagree Agree

1 2 3 4 5

DOI: 10.1057/9781137303028

The workshops draw out participants' disciplinary subconscious so that they and others can see the academic assumptions and biases at work. The questionnaire functions as a prompt for improved interdisciplinary performance through greater self and mutual understanding. O'Rourke and Crowley also offer a methodological approach, but their methodology transcends itself by culminating in introspection, dialogue, and adjustment.

In his account of the nature of understanding in *Being and Time* (1927) Heidegger makes a similar point through his concept of attunement (in German, *Stimmung*). By attunement Heidegger means no mere feeling or intuition, but rather a process where we open ourselves to the particularities of a situation in order to get in sync with it. Like the tuning of a violin, we strive to 'dial in' to the individual situation. We allow the specific case to show itself in its own way. Of course we will eventually intervene, bringing our training and skills to bear upon the situation. But all thinking, and especially interdisciplinary thinking, should begin in receptivity and rhetorical nuance rather than with methodological prescriptions.

Skill at interdisciplinary work thus becomes a matter of character rather than methodology. Interdisciplinary work requires the development of a peculiar set of virtues – or perhaps better said, the development of a set of virtues to a particularly heightened degree. These include the openness to new perspectives, a willingness to admit the inadequacies of one's own point of view, to be wrong and to play the fool, and generosity in interpreting the position and motives of others. Rhetorical skill thus plays as much of a role as logic, as we adjust our diction as well as our standards to the exigencies of the moment. This isn't relativism or a situational ethics. These are adjustments that issue in a successful outcome – or not.

Of course all academic work, and human society in general, depend on such virtues. (The diminution of such virtues lies behind the sad state of so much contemporary political discourse.) But they are even more crucial in interdisciplinary work. Moreover, as Aristotle noted, the exercise of the virtues is no mere theoretical exercise: we acquire such virtues through embodying them in a practice until they become part of us. Interdisciplinary projects thus make peculiar psychological and even spiritual demands upon participants.

Now, while few will disagree with this list of virtues or the corresponding set of vices, they are likely to be greeted with a knowing smile. Talk

DOI: 10.1057/9781137303028

of 'virtues' offers a merely subjective account of human interaction. And so we return to the hard-headedness of method.

But what are the presumptions of this commitment to method? Paul Forman describes an allegiance to proceduralist methodology as one of the signature elements of disciplinarity. He notes the influence of John Rawls's *A Theory of Justice* here, the most important book in political philosophy in the past 50 years:

> Implicit in Rawls's definition of pure procedural justice is a still broader axiom that the means sanctify the ends: not only do the right means produce the right ends, and only the right means can be relied upon to produce the right ends, but, most important, the rightness of the means is the only criterion for the rightness of the ends. (Forman 2012)

This is the magic of methodism: it allows us to bracket discussion of purposes and goals and give the patina of objectivity to the outcome. Of course procedures are helpful and even necessary when dealing with a large number of cases. But proceduralism contains its own bias, in that it assumes that one standard is fairly applicable to all. This is especially problematic for interdisciplinary projects, which are often defined by their distinctive nature. Holding to a procedure where "the rightness of the means is the only criterion for the rightness of the ends" eliminates the role of judgment. And procedure that is not balanced by judgment or *phronesis* threatens to become a type of tyranny.

V Rigor in neoliberal times

> No greater service can be rendered to the history of science, at this juncture, than by relentlessly insisting upon the necessity of raising the standard of scholarship as high as possible.
>
> George Sarton, *Science*, 1918[8]

Should scholarship aim for endlessly rising standards? For that matter, is there one set of standards for scholarly excellence, or should we recognize different standards applicable to different contexts?

These are questions rarely asked. Disciplinary knowledge production determines what counts as good work by reference to standards internal to the academy. Qualities such as nuanced scholarship, originality, and mastery of the literature – summed up in the term 'rigor' – are the measure of excellence. The same point applies to 'relevance': academics accept

DOI: 10.1057/9781137303028

intradisciplinary standards for what counts as relevant research. What someone in another discipline or the larger world thinks of this work is generally of little concern.

As I noted in the previous chapter, there is evidence that this is in the process of changing. It's not that the qualities listed just above are being dismissed; they are exemplary values, as far as they go. But in the future they will be juxtaposed with another, wider set of perspectives. Academic work is increasingly evaluated from a variety of different viewpoints, a fact likely to only increase with time. The creation of assessment criteria by individual states across the US (generally, but not entirely, at the secondary level), NSF's inclusion of a broader impacts criterion in the evaluation of grant proposals, the ongoing development of a broad-based accountability culture in Western Europe and the US, and the rise of non-academic sites for knowledge production and new technologies for disseminating and evaluating knowledge – all these factors point toward the redefinition of what counts as quality knowledge, so that it includes more than disciplinary standards.

As a group, academics have ignored these developments.[9] In part this is because of a peculiar feature of 20th-century academia: the expectation of a lifetime appointment in the industry, that is, tenure. Tenure is being eliminated in full view of all, with barely a quarter of teaching positions in higher education today consisting of tenure-stream appointments. But once someone has managed to secure one of these positions amnesia reigns: they are at leisure to ignore this trend. (After all, they must have been just a little smarter or harder working than their unemployed brethren.) When these new standards of evaluation are brought to their attention academics view them as an unreasonable imposition, the intrusion of viewpoints external to academic research, and stalwartly oppose them. Then they return to their disciplinary lives.

It's a response that is at once epistemologically benighted and politically clueless. Ignoring the possibility *and* usefulness of alternative notions of rigor is a sure way to have new standards imposed on the academic community from the outside. Academics should embrace the project of identifying new perspectives – they might as well, they are coming one way or another – and treat them as fresh opportunities for research, developing case-specific, explicitly pluralist approaches to the question of what counts as quality work.

My point is likely to be misunderstood, so let me be absolutely clear. I am *not* suggesting that we dismiss, distain, or throw out the commonly

DOI: 10.1057/9781137303028

accepted standards for excellence that have come to typify academic research. These standards are fine in their place. The challenge is to *determine* their proper place today, under vastly changed social conditions, and to make room within the academy for alternative standards that do a better job of evaluating research of an inter- and transdisciplinary nature.

Questions such as, what first-cut insights can a hydrologist offer to a planning board? Or, how can a philosopher quickly, under the pressure of time, sketch out the ethical complexities embedded within climate science? Such questions combine the standard disciplinary virtues with a mixture of qualities such as timeliness, rhetorical adeptness, and attention to cost. Skills such as properly framing an ethical debate to fit the rhetorical situation one finds oneself within should count in favor rather than against one's tenure and promotion, and should be viewed as the complement of rather than as a challenge to traditional disciplinary scholarship. In addition to disciplinary work, academics should imagine what rigor looks like under inter- and transdisciplinary conditions. Such considerations could be approached as an interesting theoretical question, with the added benefits that it might help academics preserve a degree of autonomy and self-direction within their work while at the same time open up new job possibilities.[10]

These opportunities are real, but so is the danger that new inter- and transdisciplinary standards can become a stalking horse for the neoliberal agenda. In opening up our understanding of rigor to more-than-disciplinary standards we could also strike at the integrity of university life. Universities could lose their critical function, being reduced to a one-dimensional tool of marketing and the status quo. Again, to be clear: reducing universities as market entities is a category mistake. University life cannot be mapped onto the producer-consumer relation, if for no other reason that one engages in research or attends college in order to *educate* and thus *change* one's soul, proclivities, and desires. This is the difference between a liberal and a technical education. Thus what is called for is an artful balancing of attention to the needs of the larger community with the scholar's loving care. Pursuing this calls for a constant process of renegotiation. It's a tension that should not be resolved in either direction. Disciplinary and extra-disciplinary perspectives and standards need one another.

Terri Ginsburg sounds an alarm here, arguing that since the end of the Cold War the field of area studies (e.g., Russian Studies, East Asia Studies)

DOI: 10.1057/9781137303028

has lost its critical force, having now turned to more pragmatic and utilitarian focus useable by the military-industrial complex (Ginsberg 2011). Steve Fuller has also noted the ways in which applying the logic of the market to university life fundamentally misconstrues the function of universities. He has, for instance, warned that the unbundling of the Humboldtian tie between production (research) and consumption (teaching) portends the destruction of the traditional university (Fuller 2009).

Sarton's 1918 call for ever greater standards of scholarship made sense in the context of his time. But today? The limitations of disciplinary rigor go unremarked upon. In the introduction to his edited volume *The Future for Philosophy*, University of Chicago philosopher (and law professor) Brian Leiter finds the point so obvious that it can be placed within rhetorical parentheses:

> (Which "camp" of philosophy could possibly be committed to less careful analysis, less thorough argumentation?)

On Leiter's view – which summarizes conventional wisdom across the academy – there are no opportunity costs associated with rigor, no real expenses to its pursuit, financial or otherwise. And so across the 20th century, researchers were trained to embrace academic rigor as an absolute and ever-ascending value. In the larger world we constantly balance competing goods: freedom versus security; assisting others versus creating a culture of dependency; personal indulgence versus physical fitness. But not in the academy. Limits of time and cost are recognized, of course, but they are treated as unpleasant disruptions from the proper order of things rather than legitimate and even salutary aspects of a research project.

The pursuit of disciplinary rigor has led to an ever-expanding number of areas of specialization. Forman argues that the process is driven by what he calls the master concept of sociology: the "assumption of a natural and inevitable process of increasing functional differentiation of social, cultural, and occupational roles, resulting in increasingly autonomous institutions operating to ever- higher functional standards" (Forman 2012). Inevitably, this has come at the cost of broader societal relevance. Conversations are closed to anyone outside these constricting circles of expertise. Real-time research and assessment is lost as knowledge products are not produced on the timeline needed by users. But this is what has led to increasingly common calls for inter- and transdisciplinary approaches to knowledge in the first place.

DOI: 10.1057/9781137303028

The process whereby expertise undercuts democratic decision-making makes up only part of the story. Ours is a time of constant techno-scientific advance – something we seem in the main to be quite sanguine about – largely driven by knowledge creation. But there are two points worth noting about this. First, the occasions where insights in the lab lead to products in the market provides cover for a wide range of research whose extra-disciplinary value is much more dubious. The view since Vannevar Bush has been that this is simply part of the necessary serendipity of the research process – even though we can find numerous examples of use-based basic research beginning with the Manhattan Project.

Second, this is an odd way to define 'success'. There are currently 5.7 million people employed in the transportation business in the US, as truckers, taxicab drivers, UPS and Fed-Ex drivers. There are predictions that by 2025 entire fleets of driverless vehicles will be on the road. The advent of driverless vehicles will doubtless be seen as a success – although one wonders how the unemployed drivers will view this triumph. Success is defined globally, while the costs to one or another group are taken as an unfortunate fact of progress. It is a peculiar type of technological impotency: we are powerful enough to design new technologies, but not powerful enough to control their effects.

Reconsidering rigor also implies rethinking of the nature of expertise. While there is a growing literature on expertise (e.g., Crease and Selinger 2006), it has not raised the question of whether academic culture has lost its sense of proportion and may be pursuing an excessive degree of expertise. The desirability of expertise is a given; we are left with the task of mitigating its negative aspects. There has been little consideration on how to strike a balance between the sometimes competing values of technical knowledge, accessibility, cost, and time constraints.

Disciplinarity only recognizes a vertical direction to thinking. Advancement in knowledge resembles a funnel: disciplinary knowledge is often described in terms of a stovepipe, but more accurately it gets progressively narrower as one gains more education. But rather than only measuring for depth, why not integrate under the curve? Treat knowledge in terms of the space that is covered, plotting knowledge not only by the deep and narrow, but also in terms of the broad. We could then, for instance, consider the possession of three master's degrees as equivalent to a PhD, acceptable for research as well as teaching. The lack of 'state of the art' knowledge in one specific area would be compensated by an unparalleled breadth of understanding.

DOI: 10.1057/9781137303028

VI Toward a philosophy of interdisciplinarity

> Whoever wants something great must be able to limit himself.
>
> Goethe

Interdisciplinarity is derivative in nature, for we cannot truly talk of interdisciplinarity until after the creation of disciplines. Disciplines in the strong sense date from the last quarter of the 19th century, dependent on the creation of majors (post 1869, at Harvard), the existence of departments, an emphasis on the creation of new knowledge rather than the perpetuation of established truths, and the process of inter-departmental exchange: Forman notes that up through 1927, 13 of the previous 14 men hired by the Cornell Physics Department had their degrees from Cornell. In fact, we do not find talk of 'interdisciplinarity' in print until 1930 (Calhoun and Rhotan 2010).

The same point can be made concerning the philosophy of interdisciplinarity. One could describe a prehistory stretching back to Plato, and list interdisciplinarians *avant la letter* such as Descartes, Hegel, Comte, and Dilthey, 20th-century logical empiricists and the Unity of Science movement, up to and including Heidegger. But none of these thinkers' work began from the *locus classicus* of interdisciplinarity – the cultural ferment across higher education in the 1960s, which picked up speed in the reorganization of knowledge driven by the computer revolution and the Internet, which decisively shifted knowledge production away from the control of universities.

Defined in this way, the first explicitly philosophical works on interdisciplinarity consist of the essays of Erich Jantsch and others in the 1972 OECD volume and Lyotard's *The Post-Modern Condition* (1979). By the 1990s a number of writers offered a methodological treatment of interdisciplinarity, their accounts often rooted in the experience of running undergraduate programs. Interdisciplinarity was thus treated as a matter of pedagogy rather than as a culture-wide crisis in knowledge production.

If we need a philosophy of interdisciplinarity today, as I believe we do, this does not mean that we require professional philosophers in order to think about interdisciplinarity. Academic philosophers, having formed their own discipline, are entertained by their own problems, and so are usually latecomers in considering new cultural developments.

DOI: 10.1057/9781137303028

The first compendium on the topic by philosophers occurred only in 2012, in a special issue of the journal *Synthese* (Hoffman, Schmidt, and Nersessian 2013). But we do need wide-ranging, philosophical critiques that frame issues of interdisciplinarity in terms of larger questions of the future of knowledge and culture.[11] So, for instance, the transhumanist movement's utopian vision of the future of knowledge is likely to become more prominent. It already constitutes one of the most portentous, if not dangerous, manifestations of the modernist spirit present in society.

If we then ask, what is interdisciplinarity? We find both negative and positive answers. Negatively, the rise in interdisciplinarity means the end of modernity: the breakdown in the separation of the public spheres and the isolation of one kind of knowledge – scientific – from other types of discourse. It also portends the end of disciplinary autonomy, as knowledge becomes a common possession of society at large, most obviously via the spread of higher education and the Internet.

Again negatively, interdisciplinarity is charged with politicizing knowledge, with being insufficiently rigorous, and thereby with undermining academic authority. All of these charges are true. But the answer is not to seek to re-depoliticize knowledge, or to labor to become more rigorous, or find ways to better prop up academic authority. The social conditions underlying those points are fading or already gone.

Positively framed, interdisciplinary research is research that recognizes limits – to people's capacity for understanding, to time and money, and to research itself. Now, this may sound like a negative point as well. But that is a reflection of the fact that modernity has been defined by its disdain for limits. Capitalism is an economic system predicated on the overcoming of every limit. Technology assumes that there is always another, better device around the corner. And knowledge culture, rejecting the idea of a perennial philosophy, constantly seeks more data, information, and knowledge.

In the final analysis interdisciplinarity is an environmental concept. Properly understood it challenges and disrupts the culture of infinity, whether this is the infinity of material possessions, desires, or knowledge. A post-modern age – if we are able to get there – should likewise be framed in terms of an ecological understanding of things where sustainability will be the watchword. But then the question becomes what sustainability means for the academy.

DOI: 10.1057/9781137303028

Notes

1. J. Britt Holbrook criticizes the identification of interdisciplinarity with integration in Holbrook 2013.
2. One can learn a bit about the Rock Canyon Anticline, at http://www.earth.northwestern.edu/research/sageman/orbit.html.
3. Katri Huutoniemi, in the forthcoming *Transdisciplinary Sustainability Studies: A Heuristic Approach,* argues along analogous lines, when she speaks of turning away from methodological thinking and treating research in an ecological and situational fashion. See Chapter 4 of this book.
4. An earlier version of this essay has been published in Peter Weingart, Britta Padberg (eds.), 2014. *University Experiments in Interdisciplinarity: Obstacles and Opportunities.* Reihe Science Studies.
5. I considered including herein a case study from my (and our) own work in inter and transdisciplinarity. I chose not to, because I would not want to cite *one* case study, thus making it exemplary, while to explore several case studies would mangle the structure of this book. For insight into our case studies see Briggle (2012), Frodeman and Briggle (2012), Holbrook and Frodeman (2011), and Frodeman (2003).
6. Although the work of Steve Fuller on the subject goes back some 20 years (e.g., Fuller 1993/2004).
7. Allen Repko (2008), a colleague of Newell's with similar views, speaks of "delineating a step-by-step research process that is based in relevant scholarly literature" and of providing "an easy-to-follow checklist so students can evaluate previous research."
8. Quoted in Forman (2012), p. 62.
9. There is some evidence of movement. While most scientometricians focus on bibliometrics, that is, citation counts such as the H Index and Journal Impact Factors, there is a growing group which has sought to develop alternative systems of metrics, what are sometimes called Altmetrics. See Holbrook *et al.* (2013). Similarly, Lane and Rogers (2011) note the development of different levels of rigor for different contexts.
10. To reiterate a point made above: the Harvards and Stanfords of the world may well remain untouched by such concerns. But for the vast majority of publically supported institutions these issues are likely to become very real.
11. The closest that we have come to a philosophy of interdisciplinarity may be found within the field of social epistemology, especially in its Fullerian manifestation.

Bibliography

Ackerman, T.F., 1980. "What Bioethics Should Be," *The Journal of Medicine and Philosophy,* vol. 5, pp. 260–275.

DOI: 10.1057/9781137303028

Babich, Babette, 2013. "On the Analytic-Continental Divide in Philosophy: Nietzsche's Lying Truth, Heidegger's Speaking Language, and Philosophy," Articles and Chapters in Academic Book Collections. Paper 6, http://fordham.bepress.com/phil_babich/6.

Bammer, Gabriele, 2013. *Disciplining Interdisciplinarity: Integration and Implementation Sciences for Researching Complex Real-World Problems.* Australian National University E Press.

Briggle, Adam, 2012. "How Can We Assess Fracking's Consequences without Being Influenced by Big Business?" *Slate*, July 6.

Calhoun, Craig and Rhotan, Diana, 2010. "Integrating the social sciences: theoretical knowledge, methodological tools, and practical applications," in Frodeman *et al.* (eds), *Oxford Handbook of Interdisciplinarity,* pp. 103–118

Callahan, Daniel, 1973. "Bioethics as a Discipline," *The Hastings Center Studies,* vol. 1, pp. 66–73. Reprinted in *Bioethics: An Introduction to the History, Methods, and Practice.* Jecker, N.S., Jonsen, A.R., and Pearlman, R.A. eds., 2007 (London: Jones and Bartlett), pp. 17–22.

Crease, Robert and Selinger, Evan, 2006. *The Philosophy of Expertise.* Columbia University Press.

Foltz Bruce V., 2000. "Inhabitation and orientation: science beyond disenchantment," in Robert Frodeman (ed.), *Earth Matters*: The Earth Sciences, Philosophy, and the Claims of Community, New York: Prentice Hall.

Frodeman, Robert, 2003. *Geo-Logic: Breaking Ground between Philosophy and the Earth Sciences.* State University of New York Press.

Foreman, Paul, 2012. "On the Historical Forms of Knowledge Production and Curation: Modernity Entailed Disciplinarity, Postmodernity Entails Antidisciplinarity," *Osiris*, 27: 56–97.

Frodeman, Robert and Briggle, Adam, 2012. "The Dedisciplining of Peer Review," *Minerva,* vol. 50, no. 1: 3–19.

Frodeman, Robert and Mitcham, Carl, 2007. "New Directions in Interdisciplinarity: Broad, Deep, and Critical," Bulletin of Science, Technology, and Society, vol. 27, no. 6, Fall: 506–514.

Fuller, Steve, 1993. *Philosophy, Rhetoric, and the End of Knowledge,* Routledge. Reissued in 2004 by Fuller, Steve, and Collier, James, Lawrence Erlbaum Associates.

Fuller, Steve, 2009. *The Sociology of Intellectual Life: The Career of the Mind in and around the Academy.* London: Sage Publishing.

DOI: 10.1057/9781137303028

Ginsberg, Terri, 2011. "Contemporary Interdisciplinary Studies and the Ideology of Neoliberal Expansion," *Arab Studies Quarterly*, vol. 33, no. 3–4 (Summer/Fall 2011): 143–152.

Hargrove, Eugene, 2003. "What Went Wrong? Who's to Blame?" *Environmental Ethics*, vol. 25: 38–39.

Heidegger, Martin, 1927/1962. *Being and Time.* trans. John Macquarrie and Edward Robinson, Harper Perennial Modern Classics.

Heidegger, Martin, 1934/1968. "What Is Called Thinking", trans. Fred D. Wieck and J. Glenn Gray, New York: Harper, 1968, p. 130; quoted in Babette.

Heidegger, Martin, 1943. "On the Essence of Truth," David Farrell Krell (ed.), *Basic Writings.* Harper Perennial Modern Classics.

Hoffmann, M.H.G., Schmidt, J.C., and Nersessian, N. 2013. "Philosophy of and as Interdisciplinarity," *Synthese*, Special Issue, vol. 190, no. 11.

Holbrook, J. Britt, 2013. "What Is Interdisciplinary Communication? Reflections on the Very Idea of Disciplinary Integration," *Synthese,* vol. 190, no. 11: 1865–1879.

Holbrook, J. Britt *et al.*, 2013. "Research Impact: We Need Negative Metrics Too," *Nature*, vol. 497, no. 439, 23 May.

Holbrook, J. Britt and Frodeman, Robert, 2011. "Peer Review and the *Ex Ante* Assessment of Societal Impacts," *Research Evaluation,* vol. 20, no. 3, September: 239–246.

Huutoniemi, Katri, "Introduction," in Huutoniemi, Katri and Tapio, Petri (eds), forthcoming. *Transdisciplinary Sustainability Studies: A Heuristic Approach.* London: Routledge.

Jacobs, Jerry A. and Frickel, Scott, 2009. "Interdisciplinarity: A Critical Assessment," *Annual Review of Sociology,* vol. 35: 43–65.

Jasanoff, Sheila, 2010. "A field of its own: the emergence of science and technology studies," in Frodeman, R., Klein, J.T., and Mitcham, C. (eds), *The Oxford Handbook of Interdisciplinarity.* Oxford University Press, pp. 191–205.

Kant, Immanuel, 1785/1989. *Foundations of the Metaphysics of Morals.* Pearson.

Klein, Julie Thompson, 1990. *Interdisciplinarity: History, Theory, and Practice.* Detroit: Wayne State University Press.

Klein, Julie Thompson, 1996. *Crossing Boundaries: Knowledge, Disciplinarities, and Interdisciplinarities in the Series on Knowledge: Disciplinarity and Beyond.* University Press of Virginia.

DOI: 10.1057/9781137303028

Klein, Julie Thompson, 2005. *Humanities, Culture, and Interdisciplinarity: The Changing American Academy.* State University of New York Press.

Klein, Julie Thompson and Newell, William, 1996. "Advancing interdisciplinary studies," in J.G. Gaff and J. Ratcliff (and associates) (eds), *Handbook of the Undergraduate Curriculum.* San Francisco: Jossey-Bass, pp. 393–395.

Krohn, Wolfgang, 2010. "Interdisciplinary Cases and Disciplinary Knowledge," Frodeman *et al.* (ed.), *Oxford Handbook of Interdisciplinarity,* pp. 31–38.

Lane, Joseph P. and Rogers, Juan, 2011. "Engaging National Organizations for Knowledge Translation: Comparative Studies in Knowledge Value Mapping," *Implementation Science,* vol. 6: 106.

Leiter, Brian ed., 2006. *The Future for Philosophy.* Oxford University Press.

Levi-Strauss, Claude, 1966. *The Savage Mind.* University of Chicago Press.

Lyotard, Jean-François, 1979/1984. *The Postmodern Condition: A Report on Knowledge.* University of Minnesota Press.

Newell, W.H., 2007. "Decision-making in interdisciplinary studies," in Goktug Morcol (ed.), *Handbook of Decision Making.* CRC Press.

O'Rourke, Michael and Crowley, Steven, 2012. "Philosophical Intervention and Cross-Disciplinary Science: The Story of the Toolbox Project," *Synthese,* vol. 190, no. 11: 1937–1954.

Parchman Frank, 2005. *Echoes of Fury: The 1980 Eruption of Mount St. Helens and the Lives It Changed Forever.* St Kenmore, WA: Epicenter Press.

Sarton, George, 1918. "Review of Charles Singer, Studies in the History and Method of Science," *Science,* vol. 47: 316– 319.

Schmidt, J.C. (2008). "Towards a Philosophy of Interdisciplinarity: An Attempt to Provide a Classification and Clarification," *Poiesis and Praxis,* vol. 5, no. 1: 53–71.

Weingart, Peter, 2000. "Interdisciplinarity, the paradoxical discourse," in Nico Stehr and Peter Weingart (eds), *Practising Interdisciplinarity.* University of Toronto Press.

DOI: 10.1057/9781137303028

4
Sustainability

Abstract: *This chapter explores how we can best understand the concept of transdisciplinarity – actually a more crucial term than interdisciplinarity in understanding our current situation. I frame my argument in environmental terms, arguing that sustainability should become the master trope of transdisciplinarity. This section concludes with the claim that the central element of transdisciplinarity, aka the coproduction of knowledge, implies the recognition of limits to knowledge production – necessarily a repugnant notion for the academic status quo.*

Frodeman, Robert. *Sustainable Knowledge: A Theory of Interdisciplinarity*. Basingstoke: Palgrave Macmillan, 2014.
DOI: 10.1057/9781137303028.

DOI: 10.1057/9781137303028

I Cyborg cockroaches

In November of 2013 it became possible for hobbyists to create their own cyborg, with the help of a company called RoboRoach™. For $99 kids can buy a kit containing electrodes and a tiny backpack. After immobilizing the insects via an ice-water bath, kids can superglue equipment on the antennae and the back of a cockroach and then control its movements wirelessly through a smartphone.

After a few minutes the roaches adapt to the impulses and regain their autonomy. RoboRoach claims that no lasting harm comes to the bugs, although their antennae are trimmed in the procedure. After the experiment is over kids can remove the equipment – if they are able to catch the cockroach.

RoboRoach is sold as a way to gain an understanding of human neurology: cockroach and human neurons function in similar ways. The company suggests that such early training may lead youngsters to grow up into scientists who develop cures for neurological diseases such as Parkinson's. Or perhaps they can solve the age-old problem of how to control our children.

II Transdisciplinarity as sustainability

Sustainable Knowledge offers an account of academic knowledge production, past, present, and future, in terms of four themes: disciplinarity, interdisciplinarity, transdisciplinarity, and dedisciplinarity. The previous two chapters characterized the age of disciplinarity and explored what is really at stake in the incessant talk about interdisciplinarity. This chapter turns to the question of transdisciplinarity.

There is often slippage between the technical and common uses of a term. In a nod to common practice I have often used 'interdisciplinarity' in a wide sense, to refer to the generally held, inchoate feeling that knowledge production needs to change in the face of new demands. More accurate usage would have 'interdisciplinarity' denote changes needed within the academy, 'transdisciplinarity' to efforts to move beyond university walls and toward the co-production of knowledge between academic and non-academic actors. It is the transdisciplinary moment, marking the end of the era of peer control, which threatens to truly change things.

DOI: 10.1057/9781137303028

The previous chapter concluded by framing the argument about interdisciplinarity in environmental terms. It argued that interdisciplinarity is ultimately about the recognition of limit, a notion that lies at the root of environmental thinking. Environmentalism is about many things; but one of its most fundamental points concerns the need for us to shift our desires and behavior toward the creation of a social and ecological steady state. Environmentalism thus represents a fundamental challenge to capitalism and all other philosophies of endless growth.[1]

In framing interdisciplinarity in terms of limit I was already interpreting interdisciplinarity in terms of transdisciplinarity. For when academics go out into the world they quickly find that most people have quite specific and limited needs for additional knowledge. Continuing this point, the current chapter will be thought in terms of the master trope of sustainability. To underline this point the chapter is titled 'Sustainability'.

This chapter expands the conversation about sustainability by developing the claim that an inter- and transdisciplinary era will need to make knowledge production sustainable. The epistemological regime we have been living within, that of infinite, largely *laissez faire* knowledge production, raises a variety of concerns. Additional knowledge can lead to results that are unhealthy, costly, counterproductive, unethical, and dangerous. But more simply put, in many cases there is just not that much need for additional knowledge. These points apply first to techno-scientific knowledge production, although these pathologies also show themselves within philosophy and the humanities, a theme that I will explore in the next chapter.

Suggesting that we limit knowledge production: it is hard to imagine a claim more repugnant to the academic class! I have repeatedly seen this repugnance when presenting this argument to audiences worldwide. Moreover, this revulsion is often matched with the claim that besides being abhorrent any attempt to limit knowledge production is as a practical matter obviously impossible.

The 'impossibility' argument against limiting knowledge production – what might be termed epistemological determinism – parallels an argument that one often hears concerning technology. It is said that it is impossible to stop the advance of technology. Both claims seem pretty clearly false. To take a straightforward example, continued funding for the National Institutes of Health ($31 billion annually) or the National

DOI: 10.1057/9781137303028

Science Foundation ($7 billion) is not written in the heavens, and defunding either agency would have a marked effect upon the production of knowledge. Surely some of this research would find support elsewhere; but much of it would simply fall by the wayside. The production of new knowledge would not cease, but it would slow down. Attempts to limit knowledge production, or for that matter technological development, would be only partially successful – just as they are in other areas such as in the case of laws outlawing rape or murder. And no one argues we should abandon these efforts because they are not completely effective.

In terms of repugnance, I share something of this reaction. The desire to continue to learn and grow is in many ways a laudable one. But knowledge production today has become pathological as well as unsustainable. Additional knowledge production will not end, but we do need to counterpose the desire for additional knowledge with a better recognition of the dangers of continued pell-mell knowledge production. Indeed, if one is a friend of knowledge, one should seek to protect it from its own excesses.

Steve Fuller's work provides a useful contrast here. Across a number of books (especially Fuller 2011) Fuller has sought to rethink the nature and function of the welfare state in the face of continued and accelerating knowledge production. Fuller is an advocate of pushing epistemological limits. Proposing what he calls a proactionary approach to technoscientific advance, Fuller believes the burden of proof should be on those who want to slow down technoscientific development. Evolution is now in our hands; a new version of humanity – humanity 2.0 – is inevitable, and in fact already under construction. Let's spur this development while having society indemnify unsuccessful risk takers as well as those who left behind by new developments.

In contrast, my position owes something to Edmund Burke. I understand Burke not as a reactionary but as wanting to balance progress with attention to the fragile nature of social order. The goal is not stasis but a more gentle and orderly type of progressivism that does not breed the type of reactions that now bedevil societies worldwide. A significant portion of society today feels alienated, angry about social change that they do not understand, that is being driven by incessant technoscientific advance. Scientists, technologists, and their libertarian allies have not owned up to their responsibility in spurring the fundamentalist backlash that has become commonplace worldwide.

DOI: 10.1057/9781137303028

This chapter, then, traces out the path from transdisciplinarity to sustainability and academic limit. This is a topic that, perhaps ironically, calls for a book-length treatment all its own. But what I seek here is to place these points within the overall arc of my larger argument. Hegel argued "the truth is the whole." I am trying to give a complete account of the challenges facing academic knowledge production, at least in outline.

III Transdisciplinarity in 1940s Hollywood[2]

In *Ball of Fire*, a 1941 Hollywood screwball comedy, Director Howard Hawks updates the Snow White tale. But in this case, it's the academics who are the dwarves.

A group of bachelor scholars unaffiliated with any university live together in a mansion in New York. They've dedicated their lives to the creation of an encyclopedia. A rich benefactress, following the wishes of her late father, funds the project. At the start of the movie the benefactress visits and encourages the academics to wrap up their work. It has been years, after all, and they have only reached the middle of the letter S (they are currently working on 'sex').

Hawks and screenwriter Billy Wilder extract much merriment from the conflict. A linguist (Gary Cooper) is the leader of the group. His hunger to understand the newest hip lingo leads him to a boogie-woogie joint. There he witnesses the performance of burlesque dancer Sugarpuss O'Shea (Barbara Stanwyck, who is a salacious combination of Snow White and the Evil Queen). Sugarpuss' non-stop patter ('shove in your clutch, mister') captivates the linguist. It turns out that she is also a gangster's moll who the police want to question about her boyfriend's alibi. Needing a place to hide out, she pops into the scholars' mansion for an unexpected stay.

The upshot, predictably, is chaos. Academic order is overturned by a flood of unrequited desire. Sugarpuss, the embodiment of the erotic principle, sets fire to the dry-as-dust logic of the scholars. But she does more than simply wreak havoc. She also gives focus and direction to the gentlemen's lives. Stanwyck provides an end to the endless cycle of scholarship: rather than finishing the encyclopedia, the scholars oversee the marriage of the linguistics professor and the dancer.

DOI: 10.1057/9781137303028

Hawks's account, while fulfilling the conventions of Hollywood cinema, still manages to capture essential points about the mid-20th-century academy – the focus on rigor above every other value (such as social relevance, cost, or timeliness), the zealous protection of academic autonomy, and the academic and societal contradictions that inevitably result. Like the world of Hawks' scholars, the 20th-century academy was self-governed. Academics largely defied attempts by non-academics to influence their work, viewing this as by its very nature constituting unwarranted interference.

The result was disciplinary knowledge production, the centrifugal spinning of more and more knowledge within smaller and smaller domains. *Ball of Fire* gives us an image of disciplinary knowledge circa 1940 and the contradictions it is subject to. It also highlights how academics could be drawn into transdisciplinary relations – via the erotic impulse. This is not simply a question of sexual attraction. In ancient Greek philosophy *eros* denotes the principle of desire in general – certainly including sex, but also the desire for truth, justice, or for that matter to know God. During World War II, for instance, the Nazi threat drove large numbers of physicists to turn their research in physics toward serving the war effort. One result was the Manhattan Project which produced the first nuclear weapons.

By the erotic moment, then, I mean the point where knowledge becomes focused and directed, where there is goal or end (in Greek, *telos*) to knowledge production. Note that 'end' here functions in two senses – not only as purpose, but also as conclusion or completion to knowledge production. This is in contrast to disciplinary knowledge production, which rather than having an end is only motivated by a free-form curiosity that is essentially infinite in nature. Disciplinary knowledge endlessly proliferates in all directions – a fact that has been celebrated as a virtue when it was thought about at all. So, for instance, in academic philosophy there are approximately 200 academic journals in English and in the subfield of logic, 60 journals. Similar numbers can be found throughout the humanities and sciences.

All this disciplinary knowledge production raises a question that is rarely asked: who is the audience for this mass of material? What demand is this supply an answer to? What are the costs – economic, social, and ethical – of such proliferation? And is this disciplinary knowledge production sustainable?

DOI: 10.1057/9781137303028

IV Disciplinarity and the problem of big data

As noted in the first chapter, we live in the midst of an explosion of knowledge. This flood is mostly viewed as a benign social fact, at least from the production side. As strange asymmetry reigns: the positive outcomes of basic research justify its continued funding, but when knowledge has negative consequences knowledge producers claim that their insights and inventions are essentially discrete from social processes. Nuclear bombs don't kill people; people kill people.

There are some complaints about all this information. David Shenk has argued that we live in the midst of data smog. Society has shifted from information scarcity to information abundance, with the result that "attention gets diverted (sometimes dangerously so); conversations and trains-of-thought interrupted; skepticism short-circuited; stillness and silence all but eliminated" (Shenk 2007). In a similar vein Nicholas Carr is concerned with whether Google is making us stupid. He wonders if the constant distractions and opportunities of instant searches and hotlinks are destroying our ability to concentrate and contemplate (Carr 2008). Both authors grant the positive aspects of the information revolution, but feel that we have lost our balance in evaluating the place of information in our lives.

Nor are the costs of continuous knowledge production limited to acts of commission. As Don Swanson noted in 1986, there is also a vast amount of undiscovered public knowledge lying about, insights to be had and discoveries to be made by connecting knowledge across different groups and disciplines. Swanson, a library scientist at the University of Chicago, used the case of Raynaud's Syndrome to show that significant discoveries can be made by examining, and connecting, existing knowledge across disparate fields (Swanson 1986). We waste what we already know in the mad rush to know more. Of course, this happens within as well as across disciplines. A friend of mine, a planetary scientist, has a two-foot tower of CDs in his office gathering dust, data from old missions to Venus. It's information that has never been analyzed – and is never likely to be, as ever greater amounts of data arrive from newer missions.

Thus the problem of big data. Of course, it is possible to create computer programs that can scan such data for interesting patterns. But computer programs do not have the interpretive skills of humans, and thus are liable to miss what is most significant. In one of the most important discoveries of 20th-century science, Joseph Farman spotted an anomaly

DOI: 10.1057/9781137303028

in atmospheric data from the South Pole. NASA satellites had detected an ozone hole forming over the pole, but the satellite's data-analysis software was programmed to discard the data as anomalous. Farman looked into the data on his own – despite attempts by superiors to assign him to other tasks. His (and colleagues) 1985 paper in *Nature* showed that ozone levels over Antarctica had fallen by about 40% from 1975 to 1984 (Farman, Gardiner, and Shanklin 1985). The US Environmental Protection Agency (EPA) concluded that such a decrease in solar protection could cause an additional 40 million cases of skin cancer (Vitello 2013). The eventual result was the Montreal Protocol, an international treaty designed to phase out the production of substances that deplete the ozone layer.

But note that continued, increasing, and largely *laissez faire* knowledge production raises dangers as well as inefficiencies. Once, at a conference at Columbia University, I asked transhumanist thinker Ray Kurzweil whether he was concerned with the negative effects of powerful new technologies resulting from scientific discoveries. He replied that the positive aspects of new knowledge would "outweigh by a 1000 to 1" any negative effects. But what if the 1 is powerful enough to provoke a catastrophe? Would a 1000 wonderful outcomes matter then? (I received no reply.) As computer scientist Bill Joy notes, we might have been lucky that Ted Kaczynski, better known as the Unabomber, was a mathematician rather than a biochemist. Cool new technologies will be cold comfort if a single mad biochemist creates a designer pandemic, a scenario already imagined by Margaret Atwood in *Oryx and Crake* (2003).

Massive increases in the quantity of knowledge, matched with the enormous growth in the accessibility of knowledge, are leading to qualitative changes in knowledge culture. Old systems are breaking down. In 1900, 250 doctorates were conferred in the United States; in 2007, nearly 50,000 research PhDs were awarded (Thurgood, Golladay, and Hill 2006). This increase occurred across a period where the population of the US increased by two and a half times. A system built for endless growth is approaching steady state or even retrenchment. Technological innovations such as MOOCs could mean the elimination of scores of universities. If knowledge is power, what will be the social and political effects of what are effectively infinite amounts of information, accessible to anyone from any Wi-Fi hotspot, rather than the majority of knowledge being physically located in libraries on university campuses?

DOI: 10.1057/9781137303028

To suggest the possibility of a pause in knowledge production strikes everyone as unthinkable. Even to raise the question of slowing the train of knowledge is to evoke an inverted world populated only by cranks, atavists, and fundamentalists. It is to propose what strikes many as a deeply conservative path inimical to everyone, regardless of political or religious affiliation. Our instincts and institutions, academic and non-academic, are dedicated to continued and largely *laissez faire* knowledge production. It is the academic correlate of our economic system, which must continue to grow in order to stay healthy. This is also the logic of the cancer cell.

This rejection of limits occurs even while the continued production of knowledge clearly has a shadow side, where the growth in knowledge engenders increasing amounts of ignorance. Recent years have seen the launching of the field of agnotology, the study of ignorance (Proctor and Schiebinger 2008). Agnotologists seek to develop a taxonomy of ignorance. They highlight, for instance, the ways that different types of knowledge are delayed. Knowledge about renewable energy gets marginalized within a society devoted to fossil fuels. Or how parties seek to develop doubt about certain types of knowledge, for instance tobacco companies about the dangers of cigarette smoking or oil companies about climate change.

But these are politically determined types of ignorance. They are in that sense accidental and able to be remedied. More problematic are those types of ignorance that are epistemological and metaphysical in nature in that they are built into the very process of knowledge production. Like an expanding balloon, the very increase of knowledge can bring us in contact with increasing amounts of ignorance. So, for instance, our increasing knowledge about the complexities of the climate system means that we find variables that we previously had no knowledge of (Frodeman 2013). Additional knowledge becomes paired with the growth of ignorance.

V Transdisciplinarity and the rise of the audit culture

Hawks's portrait of academia and society – what reviewer Michael Roberts called "Snow off-white and the seven nerds" – counts as one of the last screwball comedies. The genre was soon replaced by *film noir*,

DOI: 10.1057/9781137303028

intrigue being more in keeping with the mood post-Pearl Harbor. But while *Ball of Fire* cleverly reflects some of the contradictions of 20th-century academics, it also presents college professors as unhurried, woolly-headed creatures, an image that bears little resemblance to the harried lives of academics today. (For that matter, Sugarpuss hardly represents the contemporary non-academic, who might well possess an advanced degree of her own.)

This is a way of noting that the impetus for transdisciplinarity has changed. Mid-century professors faced nothing like the audit culture that reigns over contemporary university life. The rise of neoliberal market mechanisms applied to academia means that professors are increasingly subject to a regime that makes use of all the means of the information age to discipline their behavior (Frodeman, Briggle, and Holbrook 2012). Syllabi must be publically available on university websites, student evaluations are tabulated and used in tenure decisions, and the use of bibliometrics in tenure and promotion review has become commonplace. The use of metrics of various types – G and H Indexes, journal impact factors, and even a host of altmetrics such as the number of PDF downloads – mean that the contemporary academic is monitored to a degree unimaginable to past generations.

All of this highlights the growing importance of transdisciplinarity as a defining concept of the contemporary academy. It little matters that outside a small circle few have heard of the term. The phrase is most commonly used by European researchers to describe attempts at making knowledge more socially useful, for instance by Gibbons *et al.* (1994). The term is a little outré in the United States, where it is also regularly conflated with interdisciplinarity. But the term should also, and more saliently, be seen as part of the rise of accountability metrics and the development of an audit culture. The constant monitoring of academic work is a momentous shift: rather than occasionally being charmed by Sugarpuss, academics now increasingly face the implacable demands of the bureaucratic state.

This is not to reduce transdisciplinarity to a neoliberal tool of the audit culture. The term is more multilayered than that. There is also something deeply democratic about transdisciplinary research, in that it emphasizes the academic's responsibility to produce not only for a disciplinary cohort but also for the larger community. After all, most of us are employed within public universities; it is reasonable to expect us to honor our obligations to the body politic. The challenge today is

DOI: 10.1057/9781137303028

to preserve the multilayered nature of transdisciplinarity and not let it simply turn into a synonym for economic 'return on investment'.

In any case, concerns with the future of academic knowledge production have become a hot topic. The UK's Institute for 2013 Public Policy Research report 'An Avalanche Is Coming' (Barber, Donnelly, and Rizvi 2013) summarizes the state of the debate. It notes the pressure toward the 'unbundling' of university services, where functions currently combined within individual universities – research, teaching, certification, and administration – are now being taken over by companies such as Pearson. Pearson, for instance, now provides "online solutions for programme design, marketing and recruitment, student retention and student services," making possible the elimination of whole classes of administrative employees (Ibid.: 42).

This report also sees MOOCs as likely to take over lower division courses – the Calc 101s and Phil 101s of the world – that are often taught by graduate students or adjuncts in the US. This would destroy the economic rationale that underlies many graduate programs. In many cases, universities have graduate programs not because of the existence of a viable market for PhDs in a given discipline, but because graduate students allow universities to lower their costs. Graduate students are cheap academic labor: in my own department, Philosophy and Religion Studies at the University of North Texas, graduate students teach essentially all of our lower division courses, carrying the same teaching burden as a tenure-stream professor at perhaps one-tenth the cost.

Nor would the effects of MOOCs stop there: without graduate students teaching introductory courses, departments would lose their funding for graduate stipends, and professors will not have the student clientele necessary for filling their graduate level classes in their specialty. Professors would then be squeezed between the elimination of both lower division courses and graduate courses. No wonder tenure stream appointments now make up only a quarter of teaching positions in the US.

All these are tokens of the dawning transdisciplinary age. They highlight the inadequacy of a narrow emphasis on questions of interdisciplinary and transdisciplinary processes and methodology. Rather than treating inter- and transdisciplinarity as a question of technique, both terms call for a thoroughgoing cultural and philosophical critique. They should be viewed as signposts calling for a rethinking of the functions and institutions of knowledge in the Age of Google. For much more is at stake than simply the future of the university.

DOI: 10.1057/9781137303028

And as indicated above, questions concerning transdisciplinarity and the future of knowledge production need to be framed within an environmental context. Future knowledge production needs to be seen in terms of the concept of sustainability. To once again invoke Hegel, sustainability needs to become a central term of subjective as well as objective spirit, of the life of the mind as well as of material outputs.

VI The problem with sustainability

The notion of sustainability presents us with conceptual difficulties. In fact, it teeters on the edge of being a vacant principle. Consider a common definition: sustainability consists in the long-term maintenance of well-being. Every element here is ambiguous. The best definition for 'long term' may be homespun references to seven generations, three before and three after the current one. 'Well-being' is equally slippery; questions about the concept immediately lead to debates over what counts as a necessity versus a luxury, as well as fundamentally philosophic questions about what the good life consists of. In our radically pluralist era the question may be undecidable – although given the environmental exigencies we face, this may not relieve us of the need to decide.

Researchers usually define sustainability in terms of some combination of economic, ecological, and social factors. But this passes over conceptual intricacies of the term that include hidden but quite real metaphysical and even theological elements. Taken at its word, the term evokes the goals of transhumanists who seek a human lifespan approaching immortality. Outside of this none of us are sustainable. That is, we die. The same is true of societies (hopefully across a longer term, of course). This suggests that at some point what we actually want is *unsustainability*. Just as we expect one generation in time to give way to another, societies too should shuffle off this mortal coil. In ending death we would also bring an end to birth – otherwise think of the logjam. Evolutionary processes, biophysical and cultural, imply that one generation and one societal configuration eventually gives way to another.

One could respond that this misses the point: we simply need to adjust the concept of sustainability to mean the replacement of one generation with a new one. But what then happens to the idea of 'progress'? After all, what looks like unsustainability from one perspective may be viewed as evolution and progress from another. What we want, presumably, is

DOI: 10.1057/9781137303028

resilience or relative sustainability – stability for some extended period, followed by a relatively smooth transition to a new and hopefully improved state. The question becomes a matter of what we choose to sustain, what to augment, and what to let go by the wayside.

Viewed from the most abstract and philosophical perspective, the field of sustainability studies lives between the two concepts of limit and technology. We care about sustainability because we fear that we are approaching, or have already surpassed, one or another limit, economic, social, or environmental. Technology, on the other hand, is the get out of jail free card that suggests that we might be able to invent our way out of the need to recognize any limit to our desires. This is why Julian Simon and other cornucopians claim that human imagination makes any talk of limits nonsensical, since human creativity is the ultimate renewable resource.

Such abstract points are unlikely to satisfy environmentalists. They focus on particularities such as the unsustainability of continued economic development or the continued growth in human population. The problem is that such particular concerns easily slide into one another. To see how, take the example of energy and climate change.

Just as debates over peak oil (the point of maximum petroleum extraction, after which inevitable decline sets in) were coming to a head, directional drilling and hydrological fracturing of 'tight' rock – that is, 'fracking' – redefined the question of the sustainability of hydrocarbon resources. The natural gas available via fracking is still a hydrocarbon, of course, but natural gas emits about 30% less carbon dioxide than oil, and 45% less carbon dioxide than coal, changing what counts as sustainable in terms of carbon emissions.

The United States, long a major importer of oil, is now approaching energy independence (although this is not only because of fracking; it is in part because of conservation measures such as President Obama's increasing of automobile fuel efficiency (CAFE) standards). And this is not to even consider what could constitute the next frontier carbon-wise: gas hydrates. Gas hydrates are crystalline water-based solids that resemble ice, trapped within permafrost and beneath the ocean floor. The amounts of gas hydrates – untapped as of yet – are so vast that even a partial natural or accidental release could cause immediate and catastrophic climate change.

Technological innovation, then, makes the definition of sustainability in any given area exceedingly difficult to specify. Nor is this simply a

DOI: 10.1057/9781137303028

matter of supply. For the declining price of wind and solar power may make all of our use of hydrocarbons 'sustainable' in the sense that we may in the not-too-distant future be leaving most of them in the ground.

To complicate things further, questions of abundance in one area can also raise questions of scarcity in another. In the case of hydrocarbons, renewed abundance heightens concerns about sustainability in terms of climate. Bill McKibben echoes climate scientists in noting that there are likely to be real limits to the amount of carbon we can release into the atmosphere if we hope to keep climate change to an increase of two degrees centigrade from the pre-industrial era (a generally held measure of what contemporary civilization can live with; McKibben 2012). This points up the Hegelian element present in all discussions of sustainability, where abundance leads to scarcity, and infinity in one area elicits limit in another.

Technology could change the game in still another way. Some hope to geo-engineer our way out of a climate catastrophe. This approach would treat the global climate as simply a technical challenge. Climate change could be mitigated either through the removal of carbon dioxide from the atmosphere or through changing the solar balance by releasing sulphate aerosols into the upper atmosphere or by reflecting some of the incoming solar energy away from the Earth. Neither technology is proven today, and self-consciously modifying the climate would raise a welter of social, economic, and geopolitical questions. But it remains a live possibility, and one that would upend our thinking about what is sustainable.

Finally, it's clear that these larger philosophical questions contained within the concept of sustainability also include ethical and psychological elements. An engineer might be able to identify conditions that would be economically and environmentally sustainable but which most people would find deeply undesirable – for instance, having one's lifestyle strictly controlled in terms of energy use. But even this is deeply mutable: history is filled with examples that show that humans will adapt to conditions once thought intolerable.

All of these points (which could be multiplied) highlight the problematic nature of the term sustainability. Moreover, there is at least one final conceptual element that should be noted: we will never know that we have achieved sustainability – only if, and when, we have exceeded it. We wait for the decisive event, the moment when the system breaks; but what we see is a degradation of some indices rather than a tipping point.

DOI: 10.1057/9781137303028

This is not to say that a catastrophe might not occur. Climate scientists worry about our hitting a thermodynamic ledge, a point at which carbon dioxide emissions cause a decisive shift in climate regimes. Certainly we are risking stumbling over one or another such cliff with our improvident growth in population and energy use, declining species and stressed ecosystems, and the like. But it's worth noting the many cases of predicted doom (e.g., Ehrlich's warnings in *The Population Bomb*, 1965) that have yet to occur.[3]

VII The (over) production of knowledge

None of this means that we should dismiss the concept. Such perplexities are characteristic of many basic concepts. Lovejoy famously found 66 meanings to the term 'nature' or 'natural' (Lovejoy 1935), but that has not stopped us from continuing to use the term, in many cases quite effectively. Rather, the question turns on how to make the best use of the concept of sustainability. The intuition still holds that a variety of economic and environmental systems are degrading, perhaps to the breaking point. 'Sustainability' is a marker of these concerns. But rather than treating it empirically, as if it is a testable and falsifiable concept, the term functions better as a regulative guide.

By way of analogy, consider how the precautionary principle is used. The precautionary principle states that an action or policy should not be undertaken if the consequences are uncertain and potentially dangerous to the public or the environment. In other words, the precautionary principle places the burden of proof on those considering an action, asking them to demonstrate that the proposed action will not cause harm. As a testable hypothesis the idea is hopelessly vague. There is no means for measuring what counts as sufficient certainty, or how far the burden of proof extends in a given case. Moreover, the concept is reversible. If we apply the principle to questions surrounding the development of hydraulic fracturing, we can read it as implying that we should hold off on fracking until we can demonstrate its safety. But if we see the same question from the point of view of the social and economic risks involved in ceasing oil production then the burden of proof falls on those who seek to stop the drilling.

Nonetheless, the precautionary principle can – and does – serve a useful role in environmental debates. It functions as a guide to our thinking,

DOI: 10.1057/9781137303028

inviting us to pause before acting in order to consider the possible downstream consequences of our actions. Sustainability can serve in a similar manner. The very unknowability of when we might pass a limit, in terms of climate change, ecosystem health, or resource usage should give us pause and encourage us to restrain our more frivolous desires.

If, then, the concept of sustainability is serviceable within broad limits, I propose that we add a fourth dimension to the three recognized elements of sustainability, environmental, social and economic. Invoking Hegel's distinction between objective and subjective spirit, between the things of the material world and the things of consciousness, the three external or 'objective' dimensions of sustainability need to be complemented by an internal or 'subjective' dimension: academic sustainability. The world is a seamless whole: the academy is part of the merchandizing of desire and the inculcating of the spirit of infinity. The internal sustainability of consciousness, of our knowledge and our desires, is as crucial to our future prospects as the other three.

But in what sense can the current system of academic knowledge production be thought of as unsustainable? What criteria do we have for telling whether we suffer from the overproduction of knowledge – or for that matter, of knowledge producers? Can we identify possible signs of strain? We need some type of indice, one or another measure such as rising student debt, failing colleges, unread books and journals, unemployed PhDs, neglected or misapplied knowledge, or signs of social disruption caused by new discoveries.

These are all difficult to prove categories, but we do find some indications. In addition to the points made earlier we know, for instance, that student debt in the US passed the one trillion dollar mark in 2013 (Weinberg 2013). In terms of raw amounts of data, Eric Schmidt, CEO of Google, has noted that "Every two days we create as much information as we did from the dawn of civilization up until 2003" (cited in Barber, Donnelly, and Rizvi 2013). And the credit-rating agency Moody's warns that a number of colleges are in danger of closing their doors (Troop 2013) – a sign, perhaps, of technological shifts in knowledge leading to an oversubscribed market. Meanwhile pressures on academics increase: everyone in the academy exists in a constant state of frenzy, with both opportunities and duties that cannot possibly be fulfilled. Academic life has become quite different from the reputed life of the leisured intellectual.

Within the academy, the number of submissions to scholarly journals continues to grow faster than available journal space. By one count

DOI: 10.1057/9781137303028

(Jinha 2010) there were 23,750 journal titles in 2006. Jihna sought to estimate the global volume of research since the first modern journal, *Le Journal des Sçavans*, was published in France in 1665. He calculates that sometime in 2009 we passed 50 million mark for scholarly articles, with around a million and a half peer-reviewed articles being produced in 2006 and the number continuing to grow. But such absolute numbers, while suggestive, still do not in themselves indicate that too much research is being done, any more than one trillion dollars necessarily constitutes too much student debt.

A 2009 study found that only 40% of the papers published in prominent science journals between 2002 and 2006 were cited in the first five years after publication (Bauerlein *et al.* 2010). At the same time, academic publications continue to grow at a rate of more than 3% a year, a doubling rate of every 23 years. The title of Bauerlein *et al.*'s article is "We must stop the avalanche of low-quality research," but this assumes the problem is simply one of quality. It's far from clear that the 40% that are cited constitute the best 40% of papers, given the whims of intellectual fashion and the arbitrariness of what makes it into the light of day. Bauerlein *et al.* do not raise the possibility that what we are witnessing is the general overproduction of knowledge.

There are two possible sources of pressure that could prompt the governance, limiting, or lessening of the amount of academic research: forces internal to the academic world, or one or another type of outside influence. Internally, it seems clear that very few academics have any interest in limiting knowledge production. This despite the fact that increasingly their PhD students cannot get jobs and their own books and papers go unread. Academics are evidently quite committed to pursuing their particular areas of research, and simply want to be left alone to do more of it.

This sunny embrace of endless knowledge production mystifies me. In contrast, I feel like I am drowning in knowledge, and the idea of further production is daunting. Libraries and bookstores produce a sense of anxiety: the number of books and journals to read is overwhelming, with tens of thousands more issuing from the presses each day. Moreover, there is no real criterion other than whim for selecting one book or article over another. To dive into one area rather than another becomes a willful act of blindness, when other areas are just as worthwhile and when every topic connects to others in any number of ways. The continual press of new knowledge becomes an invitation to forgetfulness, to lose the forest for the trees.

DOI: 10.1057/9781137303028

Setting aside the possibility of internal governors to knowledge production, what of external pressures? Externally, there are four reasons that the academy could face a (perhaps abrupt) shift toward something closer to an epistemic steady state: disruptive technology, neoliberalism, dangerous knowledge, and the transdisciplinary relation itself.

Disruptive technology includes all the forces that are marshaling themselves today under the sign of Google: MOOCs, Internet-based education, the open access movement, and the shifting of the center of epistemic gravity from the academy to society at large, as more and more knowledge production occurs outside the academy. Together, these threaten to upend the economic model of the academy. Of course, even if this leads to the wholesale elimination of large numbers of universities this might mean only a shift in the site of knowledge production, as it moves off campus, rather than an overall diminution. But the effects could still be profound: the university has a distinctive place within knowledge production precisely because of its distance from the proprietary interests of private firms and individuals. Entire areas of research would likely be lost, and the status of the areas that remain would profoundly change, as knowledge production would be more clearly identified with specific interests.

Neoliberalism names the shift in public philosophy over the past 40 years, the development of a new public management paradigm that seeks to govern every social institution by market relations. Under a neoliberal regime education is treated as a private rather than a public good, undercutting the financial model of state university systems, as the responsibility for funding higher education shifts from the state to the individual. The drive to apply market mechanisms to every aspect of the academy implies that areas without a clearly saleable market presence could simply disappear.

Dangerous knowledge points to the prudential concerns of the type voiced by Bill Joy. Joy worried about the dangers of personalized technoscience, what he called GNR (genetics, nanotechnology, and robotics) technologies. He saw them as so powerful that

> they can spawn whole new classes of accidents and abuses. Most dangerously, for the first time, these accidents and abuses are widely within the reach of individuals or small groups. They will not require large facilities or rare raw materials. Knowledge alone will enable the use of them.

Joy's response was to call for 'relinquishment' – the voluntary cessation of dangerous types of research. So far his call has been ignored, despite the

DOI: 10.1057/9781137303028

wide attention his article initially garnered. But the question is far from dead. Controversies such as those surrounding the invention of a new strain of the H5N1 avian influenza virus show that questions about the malign effects of knowledge will continue to surface. One major societal disruption tied to advanced research – say, the development of parahumans[4] or chimeras – could be enough to generate a societal debate about whether there should be limits to the production of knowledge.

Finally, the *transdisciplinary relation* itself, the shift toward the coproduction of knowledge where knowledge is produced in close concert with the users of that knowledge, represents perhaps the most significant challenge of all to the status quo. Most non-academics, even those who are quite highly educated, do not have a need for anywhere near as much information as is currently being produced. For most people knowledge is not an end in itself, but rather a means to other ends such as money, health, comfort, or power. Such users can be severely constricted, and constricting, in their interest in knowledge. In short, most people are looking for memos, not monographs.

Each of these possibilities implies a different type and degree of limit. No one is proposing or predicting a general cessation in the production of new knowledge. There will always be new areas that call for research. The point is not the end of new knowledge, but whether we now face a number of cases where our tacit program of infinite knowledge production will be called into question.

VIII Imagining limit

There is perhaps no idea more foreign to academic culture than the suggestion that there are, or should be, limits to knowledge production. The academy has long operated until the sign of infinity. Academics assume that every result raises additional questions, *ad infinitum*. More research is always needed, whether in subatomic physics or Milton studies.

But is this a healthy infinity? In the *Science of Logic* Hegel distinguished between good and bad infinities. For Hegel, what is bad about a bad infinity is that it has no end, in both senses of the word: no terminus and no goal. He illustrated a bad infinity with reference to mathematics, where there is always another number after the last – n+1 – as well as another number between any two numbers. In contrast, a good infinity is one that is a self-contained totality, like a circle or a Mobius strip.

DOI: 10.1057/9781137303028

The environment exemplifies both concepts. It is an infinitely renewable resource if properly managed, but is degraded or destroyed when treated in terms of a bad infinity, that is, as an infinite source of materials or as a pollution sink. The latter occurs under the assumptions of a capitalist economy, built upon endless growth, itself driven by the endless prompt of consumer desires and the endless production of new insights by academics.

In the three volumes of *Capital* Marx applies the concept of a bad infinity to the money form. His critique is also relevant to the disciplinary academy, albeit in a modified form. The modern academic knowledge machine is disciplinary in nature in that it produces primarily for its own use and by its own standards. The use of its knowledge by those outside the academy is of secondary importance (and often much less than that). The academy has institutionalized the n + 1 within the closed space of university life.

Another way to understand this point is in terms of the distinction between use and exchange value. As Marx noted, an object's use value has clear limits: one can only eat so many oranges, or use so many carpets. Beyond that point objects go to waste, implying a natural limit to production. But producing objects for exchange opens up an infinity of opportunities. In selling things you can always pile up more gold. Marx, of course, also notes that excessive exchange value leads to periodic crises of overproduction, followed by economic collapse.

The academy today finds itself in an analogous position. The 200 academic journals devoted to philosophy, and similar numbers in other fields: these exist not because the world has a need for this production, but because of the existence of an internal market peculiar to the disciplines. Indeed, the irrationality and unsustainability of the academic market is reflected in the fact that this production exists without even the requirement for consumption. How many of these articles are read by anyone other than the authors, editors, and reviewers? As noted above, studies show that more than half of all articles are not cited in the first five years after publication.[5]

Since its inception in the late 19th century, disciplinary knowledge production has embraced the logic of the n+1, assuming that the knowledge being produced would eventually be utilized in one way or another. It was not the role of the knowledge producer to coordinate discoveries with any particular 'user group'. In an era of coproduced knowledge this changes. There is now a governor, or multiple governors, on academic production.

DOI: 10.1057/9781137303028

It is hard to predict how academic institutions would be reshaped if limits to knowledge were taken seriously. Pressures could show themselves in any number of ways. Consider, as a starting point, the institutionalization of the requirement that knowledge producers provide a brief account of who (beyond disciplinary peers) might be interested in or served by a given research project. (This already occurs at the Research Councils of the UK, under what they call Pathways to Impact.) Even a simple requirement like this – say, a one page statement of how advances in bioscience could cause both medical benefits and social disturbance, or how one's study of Conrad could contribute to social justice in one or another context – could cause profound disruption across the system.

STEM (science, technology, engineering, and math) disciplines might be expected to survive through their connections to technological advance; but they would be transformed by closer ties to market concerns. As for the humanities, there are a number of possibilities. We would expect them to suffer both because of their (seeming) lack of market utility, and because the knowledge produced by humanists is often provocative rather than productive in nature. It is, after all, part of the humanist's Socratic heritage to be a provocateur. On the other hand, the wholesale elimination of programs across the humanities might cause a renewed appreciation of how central they are to the creation of a civilized society.

Humanists could protect what they cherish by proceeding along pluralist grounds. Humanists have at least three roles to play: as conservator of a common cultural legacy, as instantiation of the avant-garde, and as interpreter and translator in the borderlands between disciplines and between the disciplines and society. The first two categories have defined much of 20th-century humanities. The third, working with scientists, engineers, and policy-makers, could – and should – be seen as complementary to the first two projects. The next chapter will survey this last possibility, exploring whether there could be a non-disciplinary career track for humanists.

IX Conclusion

As seen in the reactions of the academic dwarfs in *Ball of Fire*, any talk of limits to knowledge will elicit a great deal of pushback. Some will claim that it is impossible to stop the production of knowledge. Others will argue that the very idea is dangerous or incoherent. It is a given that

DOI: 10.1057/9781137303028

our current system of knowledge production, like our economic system generally, is built on constant growth. Humanity+ enthusiasts such as Ray Kurzweil celebrate the steady acceleration of knowledge leading to technological change, and dismiss the purported dangers (moral or prudential) of continued knowledge production that concern others.

Nonetheless, the question of the possible limits to knowledge is likely to become part of our social and political conversation. At what point, for instance, will evangelicals start asking questions about advances in biomedical research? Or liberals become uncomfortable with the possibilities of the new eugenics? For that matter, do we really want to know who the winners and losers will be in an era of decisively changed climate (Frodeman 2013)?

Twenty-first-century knowledge production is likely to be defined by a struggle between competing urges for infinite and restricted knowledge production. The conversation could take many forms: We might, for instance, hear calls for limiting the production of PhDs in the humanities (the Governor of Florida has already suggested charging higher tuition for unmarketable majors) or question the purpose of dissertations on what non-academics see as arcane and useless topics of research. And we are one accident away from a serious conversation about restricting lab work on dangerous topics in bio- or nanoscience.

The age of disciplinary knowledge may be ending, but the shape of a transdisciplinary age is still unknown. What would it mean for universities to move toward an epistemological steady state? Would knowledge lose its status, or its power to persuade? Is it possible to map out a viable theoretical space other than specialized expertise, and to fashion a workable account of how much knowledge is enough (Frodeman 2011)? The next chapter explores these questions from the perspective of philosophy and the humanities.

Notes

1 This does not mean that environmentalism is opposed to all growth. The ecological economist Herman Daly notes the possibility of steady state growth, i.e., that personal and societal growth and progress are still possible within a steady state regime, through the reallocation of resources and non-consumptive types of growth.

2 Another version of this argument has been made in an essay soon appear in Huutoniemi and Tapio, forthcoming.

DOI: 10.1057/9781137303028

3 It is also true, as one reviewer noted, that predictions of doom can have the ironic effect of inducing action that prevents their occurrence.
4 Parahumans are human-animal hybrids or chimera.
5 According to the Web of Science, by 2005 48% of all publications were never cited: http://garfield.library.upenn.edu/papers/jifchicago2005.pdf.

Bibliography

Atwood, Margaret, 2003. *Oryx and Crake*. New York: Anchor Books.

Barber, Michael, Katelyn Donnelly, and Saad Rizvi, 2013. "The Avalanche Is Coming: Higher Education and the Revolution Ahead," Institute for Public Policy Research.

Bauerlein, M., Gad-el-Hak, M., Grody, W., McKelvey, B., and Trimble, S.W. (2010) "We Must Stop the Avalanche of Low-Quality Research," *Chronicle of Higher Education*, at http://chronicle.com/article/We-Must-Stop-the-Avalanche-of/65890/

Ehrlich, Paul R., 1965. *The Population Bomb*. New York: Buccaneer Books.

Farman, J.C., Gardiner, B.G., and Shanklin, J.D., 1985. "Large Losses of Total Ozone in Antarctica Reveal Seasonal ClO_x/NO_x Interaction," *Nature* 315, 16 May: 207–210.

Fuller, Steve, 2011. *Humanity 2.0: Foundations for 21st Century Social Thought*. Palgrave Macmillan.

Frodeman, Robert, 2011. "Interdisciplinary Thinking and Academic Sustainability: Managing Knowledge in an Age of Accountability," *Environmental Conservation*, June: 105–112.

Frodeman, Robert, 2013. "The Geosciences, Climate Change, and the Virtues of Ignorance," GSA Special Paper 502, Geological Society of America, pp. 145–152.

Frodeman, Robert, Briggle, Adam, and Holbrook, J. Britt, 2012. "Philosophy in the Age of Neoliberalism," *Social Epistemology*, vol. 26, no. 3 and 4.

Gibbons, Michael, Camille Limoges, Helga Nowotny, Simon Schwartzman, Peter Scott, and Martin Trow, 1994. *The New Production of Knowledge: The Dynamics of Science and Research in Contemporary Societies*. London: Sage Publications.

Hegel, G.W.F., 1831/1991. *The Science of Logic*. New York: Prometheus Books.

DOI: 10.1057/9781137303028

Huutoniemi, Katri and Tapio, Petri (eds), forthcoming. *Transdisciplinary Sustainability Studies: A Heuristic Approach.* London: Routledge.

IBM Big Data, 2013. At http://www.ibm.com/developerworks/library/bd-bigdatacloud/., accessed on September 10, 2013.

Jinha, Arif E. 2010. "Article 50 Million: An Estimate of the Number of Scholarly Articles in Existence," *Learned Publishing*, vol. 23, no. 3: 258–263.

Lovejoy, A.O. and G. Boaz, 1935. A Documentary History of Primitivism and Related Ideas. A Documentary History of Primitivism and Related Ideas. Baltimore 1935.

Marx, Karl, 1867/1992. *Capital.* UK: Penguin Books.

McKibben, Bill, 2012. "Global Warming's Terrifying New Math," *Rolling Stone*, July 19, 2012, at http://www.rollingstone.com/politics/news/global-warmings-terrifying-new-math-20120719, accessed on September 13, 2013.

Nicholas Carr, 2008. "Is Google Making Us Stupid?" *Atlantic*, July 1, 2008. http://www.theatlantic.com/magazine/archive/2008/07/is-google-making-us-stupid/306868/, accessed November 4, 2013.

Proctor, Robert and Schiebinger, Lisa eds, 2008. *Agnotology: The Making and Unmaking of Ignorance.* Stanford University Press.

Shenk, David, 2007. "The E Decade", *Slate*, July 25, at http://www.slate.com/articles/arts/culturebox/2007/07/the_e_decade.html

Swanson, Don, 1986. "Fish Oil, Raynaud's Syndrome, and Undiscovered Public Knowledge," *Perspectives in Biology and Medicine*, vol. 30, no. 1: 7–18.

Thoreau, Henry David, 1854/2008. *Walden; or, Life in the Woods.* Wilder Publications.

Thurgood, Lori, Golladay, Mary J., and Hill, Susan T., 2006. *US Doctorates in the 20th Century.* Arlington: National Science Foundation.

Troop, Don, 2013. "College's Closure Signals Problems for Others, Credit-Rating Agency Says," *Chronicle of Higher Education*, June 17, at http://chronicle.com/blogs/bottomline/colleges-closure-signals-problems-for-others-credit-rating-agency-says/

Vitello, Paul, 2013. "Joseph Farman, 82, Is Dead; Discovered Ozone Hole," *New York Times*, May 18, at http://www.nytimes.com/2013/05/19/science/earth/joseph-farman-82-is-dead-discovered-ozone-hole.html?pagewanted=all&_r=0, accessed on May 29, 2013.

Weinberg, Corey, 2013. "Federal Student-Loan Debt Crosses $1-Trillion Threshold," *Chronicle of Higher Education*, July 17.

DOI: 10.1057/9781137303028

5
Dedisciplinarity

Abstract: *This chapter argues for the dedisciplining of philosophy, and by extension the humanities. Disciplinary philosophy constitutes a category mistake: philosophy is not, or at least not exclusively, a regional ontology, as are the sciences, and never should have been treated as one. I offer an account of the evolution of this mistake, and discuss the power and relevance that a dedisciplined approach to philosophy can have to culture at wide – with the added bonus of opening up new job possibilities for philosophers. Toward that end, I propose the notions of the philosopher bureaucrat and the field philosopher as innovations necessary for refurbishing the role of the philosopher in society.*

Frodeman, Robert. *Sustainable Knowledge: A Theory of Interdisciplinarity*. Basingstoke: Palgrave Macmillan, 2014. DOI: 10.1057/9781137303028.

 DOI: 10.1057/9781137303028

> You have received a better and more thorough education that those other philosophers, and are more capable of participating in both public life and philosophy. You must, therefore, descend by turns to dwell with the rest of the city.
>
> *Republic*, Book VII

I Dedisciplining philosophy

The shift toward transdisciplinary approaches to knowledge production promises to reshape both universities and relations between academia and society. This chapter explores how philosophy can respond to this new set of circumstances. It argues that the way forward is for philosophy to dediscipline itself.

Disciplinary philosophy – which of course is not even recognized as a type or category of philosophic thought – constitutes an unnecessary limitation of the philosophical enterprise. Philosophy should not be treated solely as a discipline or regional ontology where philosophers primarily write for one another, as is the case for chemists or biologists. While there is a disciplinary dimension to philosophy, it is also something wild and untamable that should be practiced everywhere: in every precinct of the university, in the public square and the private corporation. Undergraduate and graduate programs need to make the adjustments necessary so that their students are trained to inhabit all of these spaces.

I was recently invited to speak at a large state university. During this visit I had opportunities to talk with members of the university administration. Knowing that the state had gone through hard times recently but thinking that things had taken a turn for the better, I asked about the number of new positions being advertised for the next fall. I was told that the university's college of liberal arts was going to post 64 new positions. I then asked, how many of them would be tenure track? The answer: four. Already the majority of this school's philosophy faculty is non-tenure track, living on year-to-year contracts while teaching eight courses a year.

DOI: 10.1057/9781137303028

It is far from clear that the 20th-century, exclusively disciplinary model of philosophical research will survive at more than a handful of elite schools. In what follows I respond to this situation by arguing for a pluralist approach to philosophy which matches the status quo with an alternative practice that I will call field philosophy. I emphasize that the two should be treated as complementary rather than antagonistic, and can in fact strengthen one another. But whether it is field philosophy or some other approach that comes to the fore, the status quo within philosophy increasingly appears to be unsustainable. We need a systematic program for moving philosophy beyond the walls of the disciplinary academy.

II Philosophical times, disciplinary philosophers[1]

The US military has an extensive fleet of unmanned aerial vehicles, better known as drones. From a total of less than 50 in 2002, by 2009 the services had more than 7000 unmanned aircraft, a number that continues to grow. Drones flying over Iraq, Afghanistan, and Yemen collect intelligence and target enemies, while being operated by servicemen and women sitting in air-conditioned control rooms 8000 miles away in Nevada.

Drone technology highlights the unconventional nature of modern warfare. The cyber warrior who drops a smart bomb on a target on the other side of the planet – when she goes to her daughter's soccer game that evening, does she count as a combatant? What of the computer engineer in Palo Alto who designed the software that makes such precision bombing possible? Has science and technology changed the definitions of civilian and soldier? If an enemy combatant comes to the US and kills the software designer in his university office, is this an act of terrorism, or of war?

Technoscientific advances are perhaps the single greatest generator of such questions today. They are changing our political system, our economic relations, the condition of the natural environment, even our own minds and bodies (Allenby and Sarewitz, 2011; Fuller 2011). Instant access to information (Google), frictionless global communication (Skype, Facetime), and multiplying individually sourced narratives (Web 2.0) are reshaping the nature of knowledge, which in turn is remaking every other aspect of society.

The problems being broached are often deeply philosophical in nature. They raise new questions about what it means to be human, how we relate

DOI: 10.1057/9781137303028

to one another, and what our fate will be. They challenge the seemingly stable configurations that have governed our lives, such as the distinction between civilian and soldier, or human and machine. But these philosophical points often lie buried beneath conventional attitudes that view nearly every question as being either scientific or economic in nature. Given the widespread bias that philosophy is irrelevant to practical matters, the philosophical elements of our lives often need to be excavated in order to be appreciated.

Public discourse, then, needs an infusion of philosophical reflection. Moreover, in a time of tightening budgets as well as demands that the academy offer a better account of its activities, these issues present philosophers with a chance to make the case for their significance to society.

But let's be clear about the point being made here. I am not arguing that philosophers should conduct more research in social and political philosophy. Nor am I calling for philosophers to once again take on the role of public intellectual. Nor, finally, am I claiming that philosophers and humanists generally should become activists.

Any or all these developments would be welcome, but my point operates at a different level: philosophers need to become active participants in, rather than only commentators on or protesters of, ongoing policy debates. This means working alongside and at the project level with scientists, engineers, policy-makers, and public agencies. Philosophers, and humanists generally, need to get out of the study and into the field (Frodeman 2010).

This chapter develops these claims. Viewing 20th-century philosophy as unreasonably disciplinary in nature, it calls for the dedisciplining of philosophy.[2]

Central to the project of dedisciplining philosophy is treating the institutional aspects of philosophy as a philosophic issue. The institutional status of philosophy – that is, the fact that there are philosophical presumptions embedded within its functioning as a discipline – constitutes the great blind spot of 20th- and now 21st-century philosophy. It has affected what has counted as quality thinking in both form and content. And it is largely what has led philosophy, the most pertinent of subjects, to become a synonym for woolgathering and irrelevance. Put philosophers inside a philosophy department, and it is very hard to ever get them back out again.

Philosophers see themselves as being the university's critical intelligence par excellence. But they have simply punted on the question of the

DOI: 10.1057/9781137303028

institutional expression of philosophy, and the relation of philosophy to disciplinarity. There are four sets of questions that should be explored:

1. Why are philosophers housed in philosophy departments? Why aren't they spread across the various departments and units of the university? Doesn't their ghettoization imply that philosophy is simply another regional ontology on par with geology and chemistry?
2. Why is the only real job possibility for someone with a PhD in philosophy an academic position in a philosophy department? Are there no issues in other departments, and in the public and private sectors that require significant philosophical thought, and a philosopher's actual presence?
3. Why are PhDs in philosophy the only persons considered competent to judge philosophical work? Isn't everyone a potential peer when the subject is philosophy?
4. How do standards for what counts as quality philosophical work change when we expand the space of philosophy beyond the realm of the disciplinary? That is, not only do we need a philosophy of interdisciplinarity that asks questions about our standards of method and rigor for inter- and transdisciplinary projects (Chapter 2). The same questions need to be asked about philosophical research itself.

This point can also be made in terms of scale. Philosophers today are accustomed to living professionally within disciplinary precincts – the solitary philosopher writing in the study, and lecturing in the classroom, hoping for influence somewhere, somehow, in the future. But insofar as they have ventured outside, philosophers have mainly functioned at one of two scales. They have had hopes of operating on the macro scale, as a modern version of the philosopher king, called in for a chat with the university president (or Obama himself). In the 20th century this view has sometimes been expressed in terms of philosophers functioning as public intellectuals. Or philosophers have operated on the micro scale. This could involve teaching inmates at a prison, or perhaps taking part in political actions or working in a local soup kitchen, an activist's approach to public issues.

Both roles are worthy ones and should be encouraged. But the bulk of practical (and potentially employable) opportunities for philosophers lie at a scale between these two extremes. The truly massive opportunity lies at developing a new social role, what might be called the philosopher

DOI: 10.1057/9781137303028

bureaucrat. The failure to spot this middle space between the philosopher *qua* public intellectual and the philosophy *qua* activist has left philosophy on the sidelines of a thousand debates that are deeply philosophical in nature.

Now, this term may be off-putting. Few grow up dreaming of becoming a bureaucrat. But ours is a global age where most people work for, and our lives are governed by, large bureaucratic institutions. Such institutions – state, federal, and international agencies; local, regional, state, national, and international business organizations and NGOs; and universities – are where most of the action is. (The irony about complaints with the term 'philosopher bureaucrat' is that academics already live thoroughly bureaucratic lives.)

Unless we want to consign philosophy to the diminishing realm of academia, or to the uncertain possibilities of freelance entrepreneurship, we need to create spaces for philosophy within the (other, non-university) bureaucratic realms that surround us – areas which are already wrestling, often badly, with philosophic questions of one type or another. 'Philosopher bureaucrats' – or by whatever other name you prefer – thus represent a signal opportunity: working at the institutional level, in the public and private sectors, on live problems, in real time (Frodeman, Briggle, and Holbrook 2012).

This chapter, then, explores the crucial missing element of 20th-(and now 21st-) century philosophy. Whether through insufficient creativity or failure of will, 20th-century philosophy misapplied itself.[3] Philosophy across the 20th century operated as a regional ontology in principle no different from any other discipline across the academy. In so doing it produced some interesting philosophical work. But it also missed crucial opportunities, such as being the integrating element across the disciplines, and engaging in a policy-oriented philosophical practice. One example of these missed opportunities is the near total lack of work in the area of the philosophy of policy (but see Frodeman and Mitcham 2004).

III The history of a prejudice

The path by which a disciplinary-oriented, socially disengaged philosophy came to be the unquestioned standard for the field forms part of the larger story of changes across 19th-century society and culture. In

DOI: 10.1057/9781137303028

brief, it is the tale of philosophy being pushed from the center of high culture, a result driven by changes in technology and the resulting alterations across American life. Changing societal needs caused innovations within the production of knowledge, particularly the development of the natural and social sciences out of natural and moral philosophy. In the end, however, the marginalization of philosophy was a process that philosophers themselves were complicit in.

From the founding of Harvard College in 1636 until the late 19th century, an American college education was rooted in the liberal arts. Men attended college in order to prepare for a role in the upper echelons of society. Vocational training, whether in medicine, the church, or the law, was preparation for the work God called one to do. It was fundamentally philosophical and theological rather than technical in nature (as expressed by today's much different sense of vocational training as 'Vo-Tech'). Scientific training, which did not include experiments, was rooted in natural philosophy and theology, speculative fields that sought to reveal God's purpose by studying the world he created.

Up through the first seven decades of the 19th century a college education consisted of a standard curriculum with neither majors nor electives. Class work culminated in a course in moral or mental philosophy, often taught by the college president. Such courses would often address questions of social life, including the particularities of political economy and philosophical psychology. Such coursework had practical value at a time when only a tiny percentage of American youth (less than 1%) attended college, and college graduates could expect to play a leadership role in a knowledge-limited, agricultural society.

This type of education became increasingly anachronistic in post-Civil War America. The nation was industrializing, cities were growing – in 1880, 75% of urban inhabitants were foreign born or the children of foreign born – and railroads and telegraphs were creating a national market for goods and information. Both public and private sectors needed people trained in science, business, and the mechanical arts. In the midst of the Civil War, Congress used the Morrill Act (1862) to prompt states to establish 'land grant' colleges – the first were Michigan State and Penn State – that would focus on the teaching of science, agriculture, and engineering. These schools may have been dismissed as 'cow colleges' by those at elite institutions, but this did not hinder the latter's movement in the same direction: pushed by alumni and trustees, Dartmouth founded the Thayer School of Engineering in 1867, Columbia established

DOI: 10.1057/9781137303028

the Mechanical Engineering Department in 1897, and Harvard created its Business School in 1908.[4]

The social conditions of post-Civil War America also led to demands for reform that affected higher education. The 'social question' of the late 19th century – the tawdriness of the Gilded Age, the Boss politics, the squalor of the cities, the growing problems of poverty, prostitution, disease, and injustice – spurred the creation of the social sciences. But at the very moment that the five social sciences (political science, sociology, economics, psychology, and anthropology) were formed around questions of pressing social need, philosophy and what became known as the humanities moved in the opposite direction. Philosophy increasingly became the repository of classicism and eternal values.

The influence of Darwin on philosophy was profound: *The Origin of Species* (1859) killed off natural theology among the educated classes, while leaving a great deal of unfocused theological energy looking for a place to settle. At the same time thousands of Americans were traveling to Germany for an education; many returned home impressed with the German university system and with an interest in German Idealism. The result was that philosophy came to combine theological inclinations with the abstruse thought of Hegel and other German Idealists.

There was a class element at work as well: the rise of the natural and social sciences threatened to displace the earlier social elite, who attended places such as Harvard College before living a gentrified life as a member of the clergy, law, or medicine. With the rise of industry and the need for new competencies in science, engineering, and business this cultural elite was threatened with extinction. The path to a PhD in the humanities provided a way to refurbish the status of a humanistic education. Not many chose to pursue advanced degrees, of course, but support for the humanities and the rigors of a PhD provided an overlay of democratic openness and intellectual merit to their elite status. This was true even though it was not likely that the sons of the poor or even the middle class would have sufficient leisure time to devote to the years of study necessary for the PhD.

The overall result was that philosophy in America in 1900 was in a confused and marginal condition. The natural sciences were flourishing, and the social sciences were successfully establishing themselves within the academy. Philosophers were unsure of their place within the university. When the American Psychological Association was formed in 1892 it included a sizable segment of philosophers. But by the end of

DOI: 10.1057/9781137303028

the century psychologists had found philosophers to be insufficiently scientific, and expelled them.

In response, in 1901 philosophers formed the American Philosophical Association. This second APA faced the question of how to define the professional status of philosophers. There were a number of options. Philosophers could be the synthesizers of academic production, offering a global perspective on knowledge. They could be formalists, providing the logical undergirding for research across the academy. They could become disciplinary specialists who focused on distinctively philosophical problems in ethics, epistemology, aesthetics, and the like. They could be interdisciplinary and transdisciplinary translators who brought insights into other branches of the academy and to the world at large. They could offer a non-positivistic version of the social sciences, working in the field to address the ethics and values dimensions of societal problems. Or they could be a combination of some or all of the above.

All of these positions were potentially in play at the end of the 19th and the beginning of the 20th centuries. What wasn't debated, however, was whether this metaphilosophical question of the role of philosophy in society should become a central part of philosophical interests. There was little sense that a debate on the merits of these different positions was a necessary *philosophical* discussion that should be continuously revisited, for the health of both the profession and society.

In the event, the specialists and the formalists triumphed. This despite the complaints of as respected a thinker as William James. James served as president of both the American Psychological Association (in 1894) and the American Philosophical Association (in 1906). In his 1906 presidential address, "The Energies of Men," James offered an example of philosophy that included space for public concerns as well as those of philosophical specialists (quoted in Bordogna, 2008):

> *Every one is familiar with the phenomenon of feeling more or less alive on different days. Every one knows on any given day that there are energies slumbering in him which the incitements of that day do not call forth, but which he might display if these were greater. Most of us feel as if we lived habitually with a sort of cloud weighing on us, below our highest notch of clearness in discernment, sureness in reasoning, or firmness in deciding.*

Rather than an account of the most recent advances in academic philosophy, James's talk offered an exploration of the type of existential

DOI: 10.1057/9781137303028

challenge faced by all of us. James's talk mixed philosophy and functional psychology, technical points with general concerns.

But James's account, while noticed by the literate public, fell on deaf ears professionally. As Bordogna notes, he was unable to slow the drive toward disciplinary expertise. Philosophers abandoned what James called 'general philosophy' for an exclusive focus on 'technical philosophy'. Frustrated with the 'bald-headed and bald-hearted' younger philosophers who surrounded him, James sought to embody instead an inter- and transdisciplinary approach to philosophy, crossing borders both between disciplines and between the academy and society at large. In effect, James offered a fundamental ontology of philosophy itself. His choice, however, put him at odds with a philosophical community seeking to expel philosophical dabblers and secure its autonomy behind walls of expertise.

IV Philosophy in a time of knowledge explosion

In *Beyond Good and Evil* Nietzsche describes the dilemma facing philosophers at the end of the 19th century:

> *The dangers for a philosopher's development are indeed so manifold today that one may doubt whether this fruit can still ripen at all. The scope and the tower-building of the sciences has grown to be enormous, and with this the probability that the philosopher grows weary while still learning or allows himself to be detained somewhere to become a "specialist": – so he never attains his proper level, the height for a comprehensive look, for looking around, for looking down. Or he attains it too late, when his best time and strength are spent – or impaired, coarsened, degenerated, so that his overall value judgment does not mean much anymore. It may be precisely the sensitivity of his intellectual conscience that leads him to delay somewhere along the way and to be late: he is afraid of the seduction to become a dilettante.* (Nietzsche 1886, p. 134)

These words read ironically today. At least in terms of information, the difficulties faced by anyone who seeks the "height for a comprehensive look" have grown by orders of magnitude. Nietzsche is asking, isn't this situation itself a philosophic problem? Can a comprehensive look count for something more than dilettantism? How might a philosopher go about combining the wisdom of the old with the eros of the young? Are there strategies that we could devise to address these concerns?

DOI: 10.1057/9781137303028

Nietzsche's questions, however, elicited little response within the philosophical literature. Few thought to offer a philosophical defense of the idea that part of the task of the philosopher was to be a generalist, with a different mission than the regional ontologists and 'nook-dwellers' who inhabit the rest of the academy. When such a suggestion was made it was quickly dismissed as 'Weltanschauung philosophy' and the over-exuberance of youngsters inspired by the existentialists. Nor did anyone think to argue that the idea of the Renaissance man could be amended in light of the changed epistemic circumstances of the 20th (and now 21st) century.

Rather than exploring the theoretical implications of philosophy being a fundamental rather than regional ontology, the philosophical community responded to the massive growth of knowledge by going disciplinary: embracing Adam Smith's division of labor, and defining standards of excellence in terms of peer-based, internal criteria. Philosophical research consisted of the work of specialists – intricate analyses that made incremental additions to the literature within one or another philosophical subfield, written in prose understandable only to disciplinary (and sub-disciplinary) peers.

Consider a sample of titles (all since 2005) from three prominent journals:

Philosophical Studies

- Exemplarization and Self-Presentation: Lehrer and Meinong on Consciousness
- Why Parfit Did Not Go Far Enough
- Person-Affecting Views and Saturating Counterpart Relations

Philosophical Review

- Accuracy, Chance, and the Principal Principle
- On the Supposed Inconceivability of Absent Qualia Functional Duplicates – A Reply to Tye
- Epistemic Invariantism and Speech Act Contextualism

Journal of Philosophy

- Intrasubjective Intentional Identity
- The Subtraction Argument for Metaphysical Nihilism
- Actualist Essentialism and General Possibilities

DOI: 10.1057/9781137303028

None of these titles will mean anything to the educated public, or even to a PhD outside the field of philosophy. Indeed, such titles will be obscure to philosophers outside these specific domains of research. All of these essays presuppose familiarity with a specific literature and a specialized argot – one indication of what it means for philosophy to have become a discipline.

Compare the situation to that of another field. Chemistry is a limited domain – a regional ontology – requiring technical expertise. As non-experts, we wouldn't expect to be able to make sense out of the papers in a leading journal, although we might hope that such a journal would publish a digest for the non-technically trained. Not everyone can be, or need be a chemist.

But we are all pressed by philosophical questions, personal and professional, across the course of our life. Shouldn't this fact be reflected in leading philosophy journals? Shouldn't a significant amount of the material published in professional journals in philosophy be written in a style accessible to the majority of the public?[5]

To be clear: I am not suggesting that the work of specialists should not count as philosophy. But it should be recognized for what it is: philosophy of a particular, disciplinary type.

Since its birth in ancient Greece, philosophy has existed in tension with the prejudices, idols, and demands of the community. The conflict between the philosopher and the polis has had a variety of effects. Sometimes it resulted in a violent breakdown in communication, as in the death of Socrates. On other occasions it has been the source of creativity, as (on Strauss's reading) in the creation of the dialogue form to preserve a philosopher's opportunity to speak. But while there has always been philosophic work of a technical nature, it is only in the 20th century that philosophy cut the Gordian knot and became exclusively disciplinary in nature.[6] Philosophers dealt with the explosion of knowledge and the difficulties of communicating with a wide range of audiences within democratic culture not through innovations such as the philosopher bureaucrat, but by ending active engagement with members of the community except via teaching.

To be true to its historical self-understanding, philosophy must consist of roughly equal parts internal and external thinking – in-house conversations, *and* comprehensive accounts of issues shared with those outside the disciplinary philosophical community. The two elements complement one another: disciplinary philosophy needs the open air

DOI: 10.1057/9781137303028

and wide vistas of dedisciplined philosophy; dedisciplinary philosophy is improved by the precision and care of disciplinary philosophy. But before this complementarity can be strengthened we must first recognize what we have been doing. Like Moliere's gentleman, who discovers that he had been speaking prose, philosophers need to own up to their disciplinary nature.

The un-self-aware dominance of disciplinary philosophy is indicative of a failure that is itself philosophical in nature. The dogmatic nature of 20th-century philosophy is expressed by the fact that its disciplinary nature has not been a topic of philosophical reflection.

V Philosophical peers[7]

In representative forms of government the people rule, at least intermittently, through processes such as voting, recalls, and referenda. Particularly in the American experiment everybody guards everyone else through a system of "checks and balances." But there is at least one major institution that still follows Plato's lead,[8] remaining proudly non-democratic and insisting on self-governance: the academy.

Academics have a ready explanation for self-governance: non-disciplinarians cannot properly judge their work. And so they have devised a way to evaluate themselves: peer review. Whether in articles or books, grant applications, or tenure and promotion, review by one's peers has long been the standard means of academic life.

But who counts as one's peers? The university gives a disciplinary answer to this question, with each specialty or subspecialty judging their own. And philosophy is no different: even as philosophers claim that philosophy has special relevance to everyday life, they write for and are evaluated by their disciplinary peers. A philosopher is not considered successful unless he or she contributes to peer-reviewed publications that respond to the current debate within this or that subfield.

Should philosophy have the same relation to disciplinarity as the other disciplines – the same degree of scholarship, technicality, citations, insularity, and peer control? One of the oddities of present-day philosophy is how rarely this question is asked. Go to a philosophy department with a graduate program, and sign up for a course in ancient philosophy: it is assumed that the professor will know ancient Greek and to be up on the scholarly literature in the area. Such are the signposts and gatekeepers of disciplinary

DOI: 10.1057/9781137303028

philosophic discourse. And in themselves these are not bad things. The irony is that there *was* no secondary literature for the Greek philosophers who are being taught. In fact, there was no scholarship at all in the sense that we mean it today. Philosophers were thinkers, not scholars.

This situation can be seen as partly a matter of technology; paper was expensive and reproduction of a manuscript laborious; books were rare and journals non-existent. But it is still odd to assume that Plato and Aristotle would have been *recht* scholars writing for an audience of adepts if only they'd had the *Philosopher's Index* and an Internet connection. The Greeks recognized the difference between the many and the few, and between exoteric and esoteric knowledge. But these points are not the same as embracing disciplinarity. Socrates was notorious for speaking with people from all walks of life, and when he came to be evaluated it was by a jury of his peers consisting of 500 Athenians. He may not have liked the verdict, but he did not dispute the jury's right to pass judgment.

As I noted, facility with a foreign language and a grasp of the scholarship are not in themselves bad things. But life and philosophical reflection is in many ways a zero sum game. Time spent in learning Greek and keeping up on the scholarly literature is time not spent thinking about how the insights of Plato or Aristotle or Quine can illuminate challenges faced by the US Geological Survey or one's local community. Yes, a modicum of knowledge about Plato is needed before one can use it in another context. But the question of how this modicum is defined is itself a philosophic one.

Across the long sweep of Western history we find the point repeated: Bacon, Machiavelli, Descartes, Leibniz, Locke, Marx, and Nietzsche all wrote for and sought the judgment of thoughtful people across society. One wonders how they would have looked upon what counted as philosophy in the 20th century – a highly technical, inward-looking field that valued intellectual rigor over other values such as relevance or timeliness.

Extending the idea of a peer beyond disciplinary bounds could occur in a number of different ways. We could draw non-philosophers and non-academics into the peer-review process. And we could judge philosophical prowess by a number of other criteria: publications in popular magazines or newspaper articles; hits on blogs or retweets of books and articles; media contacts; and grants awarded to conduct inter- and transdisciplinary philosophic work.

DOI: 10.1057/9781137303028

Such suggestions will run into a stiff headwind. Compare them with the only prominent ranking system in philosophy, Brian Leiter's *Philosophical Gourmet Report* (PGR). The PGR is already an exception, in that it addresses the institutional aspects of the profession. It offers a series of judgments (in the first instance for prospective graduate students) concerning the quality of departments overall and in terms of different areas of philosophy. The PGR is thus a rarity: while the topic of a thousand conference dinner conversations, questions about the institutional status of philosophy do not form part of the scholarly literature of philosophy.[9]

But what criteria does the PGR use to evaluate departments? Leiter posts the list of the philosophers whose opinions determined the results of the survey. In 2011, 500 philosophers were canvassed; a bit over 300 replied. Why these 500 philosophers, out of the 15,000 or 20,000 PhDs in philosophy employed in universities and colleges across the US? Leiter's criteria:

> Evaluators were selected with an eye to balance, in terms of area, age and educational background – though since, in all cases, the opinions of research-active faculty were sought, there was, necessarily, a large number of alumni of the top programs represented. Approximately half those surveyed were philosophers who had filled out the surveys in previous years; the other half were nominated by members of the Advisory Board, who picked research-active faculty in their fields.

How qualities such as area, age, and educational background are 'balanced' is left unstated. Nor are any of these qualities defined when it is obvious that a variety of positions can be taken on each. For instance, 'area' could include or exclude any number of research fields, which also could be divided or subdivided in a variety of ways (e.g., in terms of granularity: is the category 'ancient Greek philosophy', or 'Plato', or 'Plato's early dialogues'? The same point can be made about 'continental philosophy', a type of philosophy Leiter is well known for disdaining). We can assume that 'research active' means 'publishes a lot'; but this is also tendentious. It likely means peer reviewed and in the best journals, but what counts as a peer and which are the best journals is not even raised, much less defined. The circularity is complete with current members nominating their colleagues.

It will come as no surprise that Leiter's results have been "remarkably stable" over the years. But more to the point: Leiter not only fails to

DOI: 10.1057/9781137303028

consider what the rankings might look like if a random cross-section of employed philosophers were surveyed. He also gives no thought to how non-philosophers would rank departments, or how programs could be evaluated according to citations outside philosophy journals, by the amount of sponsored research they attract, or their degree of community involvement. Leiter's thinking about the institution of philosophy is thus remarkably disciplinary (not to say dogmatic) in nature.

These concerns – reflection on the institutional expressions of philosophy – remain a blind spot. One might think that Cohen and Dascal's 1989 collection of essays *The Institution of Philosophy* would broach questions about the institutional status of the field. But the book fails to even mention the institutional arrangements of philosophy (!). Instead, it addresses the theoretical implications of various post-modern critiques of philosophy. Similarly, in the introduction of his 2007 edited volume *The Future for Philosophy,* Leiter raises the "meta-philosophical" question of what one does when one is "*doing* philosophy." It turns out, however, that doing philosophy means "*doing* philosophy of mind, language, and science, as well as ethics, epistemology, feminist philosophy, and the history of philosophy" (emphases in the original). It's the same old stuff: new ideas, perhaps, but still exclusively directed toward the interests of a peer audience. An expanded social role for philosophy is not even contemplated.

Today we do see incipient signs of change. Prominent philosophers call for philosophy to be turned "inside out," challenging the decades-long fascination with the philosophical equivalent of an "ornamented *Quadruple Tremolo 41* with an extra trill" (Kitcher 2011). In 2010 the *New York Times* launched a blog called *The Stone* that features short philosophical essays directed toward a literate audience. A number of recent conferences have been held in the US and elsewhere on the question of the public role of philosophy. In the UK and Europe this trend is further along, exemplified by *The Philosopher's Magazine* and *HowTheLightGetsIn*, an annual philosophy and music festival in Wales. And one can point to signs of institutional change – at the School of Public Policy at Georgia Tech, where philosophers are part of an interdisciplinary faculty; at ethics institutes such as the Center for Global Ethics at George Mason University and the Rock Ethics Institute at Penn State; in the Department of Philosophy and Religion Studies at the University of North Texas, where professors with PhDs

DOI: 10.1057/9781137303028

in fields other than philosophy form part of the department; as well as at Arizona State University, which has scattered philosophers across its campus.

So far, however, this movement has not penetrated the institutional heart of the discipline – as shown in the results of a survey of philosophy programs conducted by the Center for the Study of Interdisciplinarity (Frodeman 2013). PhD programs are not being redesigned to include training in how to write NSF grants or on how to effectively work with communities. There are few, if any, undergraduate or graduate courses that train students to work at the project level with non-philosophers.[10] And there is little or no philosophical literature that explores how our standards for philosophical excellence need to adjust to different audiences and circumstances. Nietzsche's question – whether a conceptual space exists between those who "lose themselves in wretched nooks and crannies," and dilettantism – goes unexplored.

Some will claim that extending the idea of our philosophical peers to include non-philosophers will expose philosophy to the corruptions of the *demos*. Is philosophizing to become a sheer popularity contest, where philosophers are promoted based on their Klout score, or the number of Facebook 'likes' their blog posts garner? It is true that when philosophers broaden their peer group, they will lose some control over the measures used to define philosophic excellence. But this could be all to the good.

Dedisciplining philosophy entails another risk – that philosophy will become merely an instrument for an exterior set of ends. The fear here is not that abandoning disciplinary peer review will lead us into error. Instead, it is that the only alternative to value as judged by disciplinary peers is a crass utilitarianism, where philosophic value is judged by how well it advances a paymaster's outcome. One philosopher may be labeled a success for helping a racist political candidate hone his message, while another may be labeled a failure for not sufficiently fattening a corporation's bottom line. Isn't a dedisciplined philosophy actually a return to sophistry rather than to Socrates? Won't it sell its services to whoever is buying, adjusting its message to satisfy another's agenda and criteria for success? In order to survive until the turn of the 22nd century, must we sell the soul of philosophy early in the 21st?

There are two replies to such concerns. First, philosophy existed long before the 20th-century model of academic disciplinarity came to define its nature. The struggle between philosophy and sophistry is a perennial

DOI: 10.1057/9781137303028

one, and one does not necessarily sell out by writing for a larger audience – or remain pure by staying within disciplinary boundaries.

Second, as I have been arguing, disciplinary and dedisciplinary approaches to philosophy should be seen as complementary rather than antagonistic to one another. Rigor is pluralistic: the rigor of disciplinary work is different from, but neither better nor worse, than the philosophic care required to adjust one's thinking to real-world exigencies. This is a point that bioethicists have long understood. As I noted earlier, in his 1973 article "Bioethics as a Discipline," Daniel Callahan already saw that doing philosophical thinking with physicians, scientists, and other stakeholders demands "rigor ... of a different sort than that normally required for the traditional philosophical or scientific disciplines." Bioethics exists in disciplinary and in non-disciplinary forms that synergize with one another. This suggests that we need not be forced, as a matter of general principle, to choose one set of peers over another.

As universities face growing demands for academic accountability, philosophers ought to take the lead in exploring what accountability means. Otherwise we may be stuck with Dickens' Mr. Gradgrind, "Now, what I want is Facts. Teach these boys and girls nothing but Facts. Facts alone are wanted in life." A philosophical account of accountability will also require redefining the boundaries of what counts as philosophy. We ought to engage those making accountability demands just as Socrates engaged Euthyphro on piety.

VI The limits of applied philosophy

As Bruce Kuklick notes, over the past 100 years philosophers have developed increasingly sophisticated means for analyzing philosophical problems (Kuklick 2003). Much of this work has relevance to practical concerns. But this is different from engaging actual decision-making at the project level, something that philosophers have almost entirely avoided. This is not coincidental. Disciplinary excellence in philosophy naturally leads to greater distance from real-world problems. But it also raises questions about the viability of the intellectual tradition rooted in Socrates' commitment to the *polis*.

James's presidential address implicitly posed a set of metaphilosophical questions to philosophers. I hope to have made it clear that for both philosophical and pragmatic reasons, metaphilosophy today needs to

DOI: 10.1057/9781137303028

include reflection on, and practical experiments in, the relationship of philosophy with society. This means:

- working out the theoretical and practical details of how philosophers work with and write for non-philosophers;
- the pluses and minuses of housing philosophers (and other humanists) in different departments, companies, and public agencies; and
- the means for integrating philosophical case-work into the daily tasks of the public and private sectors.

Linking the theoretical and institutional aspects of our work, and expanding our philosophical home beyond the department, should cause philosophers to revise what they think about, in what venues, with what outputs, at what cost, to what standards, and for what audiences.

As philosophy became established as a discipline, part of the larger disciplinary development of the academy, the reference community for philosophic work became more clearly other philosophers, those with technical training in the field as evidenced by the PhD. There were attempts to balance technical precision with social relevance, for instance within the Vienna Circle, which were derailed by various circumstances, including World War II, untimely deaths, and the Cold War (Reisch 2005). Philosophers as varied as Bertrand Russell and John Dewey were able to successfully write for both professional and more general audiences. But the movement across the decades has generally been toward more and more disciplinary work.

A counter movement was launched in the 1960s: applied philosophy. Applied philosophy – or applied ethics, the more common phrase – is an ambivalent term. When Thales cornered the olive market 2500 years ago one could claim he was functioning as an applied philosopher. More seriously, perhaps, one could maintain that Thomas Aquinas (who wrote on marriage) and John Locke (who wrote on education, as well as toleration) qualify. By these criteria the list can be expanded to include most philosophers across the history of philosophy.

But this misses the point, in that these are pre-disciplinary examples. Applied philosophy is a distinctly 20th-century movement that arose in response to the lack of practical application of disciplinary philosophy. The question is whether or to what degree it escaped the gravitational pull of disciplinarity. On my reading it did not, because it failed to

DOI: 10.1057/9781137303028

theorize the theoretical and institutional dimensions of writing and doing philosophy for non-peers. For instance, if applied philosophy was going to successfully work with non-disciplinary audiences, faculty reward structure would have to be changed so that faculty would be given credit for the time and effort such work requires, rather than being penalized for this work, as has commonly been the case.

Applied philosophy has had a small presence within the discipline since the 1960s. Initially the practical turn on the part of philosophers was driven by a number of societal controversies over the environment, health care, and technological failures. In response, applied ethics successfully broadened what is considered properly 'philosophical'. Indeed, Stephen Toulmin argued in 1982 that "medicine saved the life of ethics," in that medical ethics and other applied fields rescued philosophy from death by irrelevance. Applied philosophy showed that real-world problems have philosophical dimensions, and that one can make interesting and important contributions to such problems acting as a philosopher. Whether we reintroduce wolves to Yellowstone Park or how we ought to allocate scarce medical resources were not questions that could be answered by instrumental forms of expertise. They raised matters of justice, first principles, epistemology, and the nature of the good—the very stuff of philosophy.

Nevertheless, the field has remained on the margins of philosophy. Institutionalization occurred late: the Society for Applied Philosophy was founded in 1982; its companion publication, *The Journal of Applied Philosophy* in 1984, while the *International Journal of Applied Philosophy* was started in 1982. In the case of more specialized journals, *Environmental Ethics*, the first journal in its field, began in 1979 and the *Hastings Center Report*, the first bioethics journal, launched publication in 1971.

Applied philosophy has made philosophical contributions that were both theoretically nuanced and relevant to the STEM disciplines and policy-makers. Bioethics stands out here as a field which has made sustained efforts to engage non-philosophers in the nitty-gritty ethical details of medical situations. In recent years there have been similar calls for social engagement from philosophers of science (e.g., Longino 2002 and the founding of the Society for the Philosophy of Science in Practice (SPSP). But overall applied philosophy has been a failure. Not by the standard of creating another subdivision of quality philosophical work

DOI: 10.1057/9781137303028

aimed at a peer group of philosophers. But a failure if its primary goal was to assist people and institutions in all walks of life with the philosophical challenges they face.

A thorough characterization of applied philosophy over the past 50 years would require a review of each of its numerous subfields, a task beyond my scope here. But consider the examples of two areas, environmental ethics and bioethics.

Environmental ethics developed in the 1970s in reaction to larger societal forces. Earth Day and the passing of the Clean Air and Clean Water Acts all predate the inauguration of the field, which can be symbolically dated by the first known course in environmental ethics, taught by Baird Callicott in 1973.

Environmental ethics took a distinctive approach to ethical questions. Thinking about ecosystem restoration or invasive species was a far cry from the question of free will and other abstractions that occupied 20th-century philosophy. Yet while researching real-world problems was unique, the way environmental ethicists went about their work was quite traditional. Environmental ethicists did not ask how the institutional and disciplinary nature of philosophy affected the practical efficacy of their work. And in remaining largely a disciplinary endeavor, environmental ethics consigned itself to the margins of both philosophy and society.

As noted above, Eugene Hargrove has repeatedly lamented the social impotency of environmental ethics. On Hargrove's account environmental ethics had been marginalized by the philosophic presumptions underlying policy analysis, whose understanding of human motivation had been underwritten by economics. Viewing humans as *homo economicus* and assuming the dominion of subjective felt preferences eliminated the need for ethical reflection. But by the 1990s environmental ethicists were giving attention to the policy element of environmental problems. Ecofeminists, environmental justice theorists, and environmental pragmatists all grew impatient with the lack of practical application of theoretical work. In *Toward Unity among Environmentalists* (1991) Bryan Norton proposed that we "think about environmentalism as a force in public policy first and to examine philosophical questions in passing."

Norton and Hargrove (1994) later distinguished between applied and practical philosophy—the former applies theoretical principles to problems, while the latter begins with real cases and seeks to insinuate

DOI: 10.1057/9781137303028

philosophic insights into these cases in a spirit of amelioration and compromise. Similarly, Light and Katz (eds) in *Environmental Pragmatism* called for a pluralist and non-reductionist approach to environmental problems that would "identify practical strategies for bridging gaps between environmental theorists, policy analysts, activists, and the public." The year 1994 also saw the publication of Van de Veer and Pierce's *The Environmental Ethics and Policy Book: Philosophy, Ecology, Economics*, with a second edition in 1997 and a third in 2003.

This literature did highlight the need to take better account of policy concerns. Its overall weakness, however, turns on its inability to carry through on its promise to offer specific insights and strategies within the context of live controversies. Thus despite the overall goal of the volume, the essays in *Environmental Pragmatism* in the main consist of applied philosophy – general, theoretical, top-down accounts of environmental questions. There are no accounts of philosophers working through case studies of particular problems – for instance, the challenges Australia faces through the importation of the European fox, the rabbit, and the house mouse. No one disputes the need for general philosophic accounts of environmental issues. But the overall thrust of the volume lies within applied rather than practical environmental ethics. In the case of environmental ethics, the message changed, but the medium – disciplinarity – remained the same. That is, few environmental philosophers take Hargrove's and Norton's distinction between applied and practical philosophy to heart.

Such 'disciplinary capture' pertains not just to environmental ethics but to most other sub-fields in applied ethics as well. In contrast, from its inception with questions about the ethics of biomedical science and technology in the early 1970s, bioethics has been quite self-conscious about its institutional status and place in the wider world e.g., Ackerman 1980, and Eckenwiler and Cohn, 2007). This is partly attributable to the fact that bioethics has emerged from and crossed over several disciplines (philosophy, medicine, law, nursing, political science, sociology, theology, and more), whereas environmental ethics was an offshoot of philosophy alone. And members of these disciplines – even before they would call themselves 'bioethicists' – have been enrolled into public policy-making on a scale that far surpasses involvement by environmental ethicists.

Thus bioethics has escaped the kind of disciplinary capture that has ensnared environmental ethics. It has, however, fallen prey to another form of disciplining, and capture. When Callahan emphasized that

DOI: 10.1057/9781137303028

bioethics should be something useful to those who face real-world problems, this was the year after the Tuskegee trials became public, and the year before the first federal-level US bioethics commission was formed. With its landmark Belmont Report, released in 1979, that commission moved in the direction called for by Callahan.

The Belmont Report outlined three principles (respect for persons, beneficence, and justice) that "provide an analytical framework that will guide the resolution of ethical problems arising from research involving human subjects." The Belmont Report seeded an approach to bioethics, 'principlism,' which contended that these three principles formed an adequate moral vocabulary for identifying and resolving any bioethical dilemma. Principlism held a monopoly on federal level bioethics commissions for nearly 30 years (see Briggle 2010). Principlism is the approach to bioethics taught at the historically formative and still influential Georgetown program. And it is the most widely debated form of bioethics, with entire books devoted to attacking (e.g., Baron 2006) and defending (Beauchamp and Childress 2008) it.

The principlist approach has helped bioethics play the role envisioned for it by Callahan, especially in the area of human subjects research. In this way it has escaped the type of disciplinary capture that marginalized environmental ethics. But bioethics has purchased its relevance at a steep price. Bioethics has been captured in another sense. It has, as John Evans (2002) noted, become "instrumentalist." For principlism substitutes a formally rational discourse for a substantively rational one. Substantive rationality debates ends and means together. By contrast, formal rationality asks whether the means employed are being maximized to achieve assumed ends.

Bioethicists settled the question of ends in advance with their principlist method. The goals of respect for persons, beneficence, and justice are classically liberal positions that walk away from substantive debates about the good life. The only questions that remain are the instrumental ones of how best to achieve them. This allows bioethicists to fit the classically liberal frame of instrumental expertise, which in turn explains their social influence. As Moreno (2005) points out, bioethicists made a compromise. They received the social power that attaches to the label of 'moral expert' while forfeiting deep questions about the ongoing technoscientific remaking of the human condition.

This makes the transdisciplinary success of bioethics as troubling as the failure of environmental philosophy. Bioethicists pass off as a neutral

DOI: 10.1057/9781137303028

expertise what is in reality a contingent way of framing moral problems. They have constrained dialogue and narrowed options without opening up these terms of constraint for debate. This has had an exclusionary effect on concerns that do not fit the supposedly neutral mold of the principlist discourse. What, for example, are we to make of claims about humans overreaching their proper sphere of action to 'play God', or about the deeper anthropological meaning of procreation? Such questions are bracketed from the conversation.

Disciplinary knowledge is knowledge defined by subject matter, tools, or methodology, *not* by the nature of the problem being faced or the use that that knowledge is being put to. Disciplinary knowledge is knowledge that severs the relation between means and end, between knowledge production and its use by those outside the academy. Principlism is still the victim of a form of disciplinary capture because it establishes a framework in advance, prior to engagement with the particulars of a problem or question. It thus isolates and confines crucial intellectual activities solely within the domain of the experts. Indeed, this is what gives them their claim to expertise; that they have settled on *the* way of doing bioethics.

In the end, bioethics took the easy path to relevance. Socrates strode out into the world naked, with no established dogma or framework – just openness to whatever conversation would arise. This requires a tolerance for loose ends and sudden twists and a knack for quick thinking – more public intellectual than a disciplinary expert. Bioethics, by contrast, armed itself with an expertise that shaped any encounter into one about autonomy, beneficence, and justice. While it appears that they are clarifying and resolving problems, they only do so by setting aside the increasingly central questions of what it is to be a person and what we should be using our autonomy *for*.

This account of two sub-fields of applied philosophy poses key questions: how does philosophy avoid disciplinary capture? Is there a way to do work that is both socially relevant *and* richly philosophical? Or do philosophers only have two equally unpalatable choices: irrelevance based in disciplinary commitments, or usefulness at the cost of disciplinary recognition?

Dedisciplining philosophy has its dangers. By moving about in the larger world, philosophy could compromise its function as social critic, or become captured by powerful interests. Or in speaking truth to power, we could be forced to drink hemlock (or more likely today,

DOI: 10.1057/9781137303028

cause a program to be defunded by an outraged state senator). But such concerns simply highlight the need and the opportunity for new philosophical work, in that each of these problems is a *philosophical* one. In any case, clinging to the status quo in the name of academic freedom and philosophical rigor is not only unsustainable, but it is also irresponsible. Philosophers, like any professional group, have an obligation to serve the community. We need to embody our own professional code of ethics.

VII Taking philosophy into the field

> *The rhizome has no beginning or end; it is always in the middle, between things, interbeing, intermezzo.*
>
> Deleuze and Guattari *A Thousand Plateaus*

There is another approach possible. Field philosophy is a problem-oriented form philosophical of practice that treat knowledge production and knowledge use as dynamically integrated. Field philosophers work at the project level on problems that have been defined by various social actors. Field philosophy begins from problems as defined by non-philosophers, and seeks to *ameliorate* situations by making piecemeal contributions to ongoing situations (Frodeman, Briggle, and Holbrook 2012). The term is crucial: field philosophy seeks to lessen problems rather than identify perfect solutions.

Field philosophers contribute to a problem by being attentive to the particulars of the case study, drawing out, assessing, and communicating whatever philosophical dimensions may be at work. Their research is addressed primarily to non-disciplinary peers in evolving contexts of use. Their disciplinary activities are mostly oriented toward sharing lessons learned in order to improve non-disciplinary contributions. Thus in addition to disciplinary criteria of success, field philosophers seek to be evaluated by their contributions to policy processes and public debates. Rather than standard bibliometrics, this requires metrics capable of capturing positive impacts on real-time decision and discourse contexts. Evaluations thus involve a pluralist and case-by-case approach where 'rigor' is defined by balancing epistemological standards with other criteria such as timeliness, cost, and relevance (Briggle and Frodeman 2011).

Field philosophy is kin to Mode 2 knowledge production, a concept originating from STS and science policy studies (see Gibbons *et al.* 1994).

DOI: 10.1057/9781137303028

Mode 2 signifies the replacement of investigator-initiated, discipline-based research with research that is problem-focused, interdisciplinary, and socially engaged. Field philosophers emphasize the untidiness of real world thinking. Thought must often grope through ill-defined circumstances. There is no determination in advance as to problem definition or the correct framing of philosophical analysis. Moreover, there is no preset understanding of what will count as an acceptable answer. Appropriate rigor is defined on a case-by-case basis involving elements of timeliness, relevance, cost, and rhetorical success.

Culture today consists of networks and interstices, held together by connections that are intermittent, transitory, and subterranean. Communication across society and the Internet is rhizomatic: links lead in all directions prompting unanticipated exchanges and synergies. Surfing the Net stands in stark contrast to the linear model of argumentation assumed as 'rigorous' by theoretical and applied philosophers.

When Deleuze and Guattari speak of rhizomatic politics they are emphasizing a subterranean approach to problems. Philosophy has always had a subterranean quality. Challenging the status quo leads to reaction; better to let insights insinuate themselves here and there. Much of the interaction that field philosophers have with other disciplines and society at large remains half-hidden and interstitial in nature – revealing concealed premises, drawing out implicit contradictions, connecting disparate insights. Field philosophy recognizes the open-endedness of philosophic work, work done in the margins and shadows. This is in contrast both to the purity of environmental ethics' disciplinary approach and to the surety of bioethics' principlism.

To sum up, there are five elements that distinguish field philosophy from disciplinary philosophy:

- Team-orientation
- Non-disciplinary audience and framing of the question
- Knowledge production in the context of use
- Contextual definition of rigor
- Alternative models of evaluating success

First, field philosophy challenges the assumption that philosophers properly work in isolation. The romantic model of the solitary genius is of decreasing relevance today. Across intellectual life work increasingly gets done in teams; the problems we face are too complex, and typically require group competence. Philosophers need to learn how to function

DOI: 10.1057/9781137303028

as part of a group, both in the sense of working with other philosophers and in clusters spanning a wide range of disciplines and dimensions of practical life.

Second is the question of audience. Applied philosophers may eventually disseminate their products to non-philosophers, perhaps by sending the resulting book or article to someone out in the larger world. But this remains *ad hoc*: the work plan and products are not designed with the specific needs of non-philosophical audiences in mind. As a result, outsiders have no sustained effect on the theorizing, in terms of things like speed of publication or the length and language of the work.

In contrast, the primary audience of field philosophy consists of a particular group of non-philosophers struggling with a specific problem. These non-philosophers define what counts as a problem and as a solution; at least initially philosophers work within that framework, consulting with them on a daily or weekly basis for months or years at a time. This means that field philosophers by design play a marginal and interstitial role in a larger process – at least at the start of the project; for over time their influence may grow.

Third, this approach implies that field philosophy sets aside the linear model of disciplinary knowledge, according to which knowledge is produced in isolation from the context of its use. Rather, field philosophers engage in a continual process of co-production of knowledge, where the context of use for that knowledge defines the space for thinking. While disciplinary peers can retain a role in these deliberations, the field philosopher recognizes a greater obligation to the needs and perspectives of his or her non-disciplinary peers. Of course reporting results back to one's disciplinary peers still has value – especially as a means for recruiting new field philosophers. But it is not the central point. Put differently, a field approach to philosophy takes seriously Husserl's "to the things themselves," venturing out to the actual problems in the world – and staying there.

Fourth, field philosophy challenges disciplinary norms for what counts as proper theorizing. The disciplinary community no longer exclusively determines what counts as the correct framing for or degree of philosophical analysis. Instead, the degree of theoretical rigor is adjusted on a case-by-case basis, defined by an individual mixture of economic, political, temporal and rhetorical constraints. Thinking retreats to the study, only to go out again into the world, over and over,

DOI: 10.1057/9781137303028

a cycling between field and study, a never ending dialogue between practice and thinking.

Finally, part of the philosophic task for field philosophers is to determine what counts as success. Evaluation becomes a deeply philosophic process. Success is in large part defined by others, since field philosophy places the greater value on helping non-philosophers work through their own problems. At the same time philosophy must reserve its right for critical judgment. If disciplinary thinking measures success by the extent to which it increases our understanding of the world, field thinking measures success by the extent to which it address the needs of others as they define them; or put differently, by one's success in changing the world.

VIII Conclusion

I've presented one or another form of this argument on many occasions and venues. The reaction on the part of audiences not comprised of philosophers is typically positive. Scientists, for instance, usually are delighted with the idea of philosophers working with them at the project level. The reaction on the part of audiences made up of philosophers is often negative. I find this divide curious, for I conceive of the points made above as constituting a defense of philosophy.

The objections on the part of philosophers generally fall into two categories. First, many react with simple disregard or disinterest. The points made here do not strike them as qualifying as a philosophical argument, but rather as a series of generalities expressing an opinion *about* philosophy. Moreover, they report being quite content with the philosophical enterprise as it is now arranged and see little need for repair.

The second reaction is more hostile. The view here is that philosophy is already under siege, and that criticisms of the profession only further arm the philistines and other enemies of philosophy.

These criticisms turn on the assumption that the status quo is sustainable. I have doubts that it is at more than a handful of elite institutions. But this fact itself highlights an important psychological barrier: every philosophy department seeks to model itself on Harvard and NYU. The idea that philosophy departments lower in prestige would redefine what counts as quality philosophical research runs up against 100 years of institutional history.

DOI: 10.1057/9781137303028

This response also reflects the assumption that metaphilosophy – philosophizing about the institutional conditions of philosophy, and what should count as the social expression of philosophy – is not a serious topic for philosophic reflection. This is of a piece with the rejection of my claim that 20th-century philosophy was disciplinary in nature. Which makes sense: if philosophy *had* changed around 1900, this would raise the possibility of creating another, more-than-disciplinary approach to the subject. And this would again threaten all the carefully constructed judgments and hierarchies that have ruled the profession.

To me it is blindingly obvious that something important happened to philosophy in America and Europe at the end of the 19th century. The invention of the first philosophy journals, the creation of departments, and a general inward turn toward professionalization marks the disciplining of philosophy. But rather than recognize this fact philosophy remains gripped by an ahistoricism that relegates anything before the *Principia Ethica* and the *Principia Mathematica* to technically deficient curiosities, and which tacitly assumes that departments are the natural and inevitable homes for philosophy. We will simply have to wait to see whether the perspectives discussed here point toward something important for 21st-century philosophy – and culture.

Notes

1 Parts of the following argument were published as Frodeman, Robert, 2013. "Philosophy Dedisciplined," Synthese, July 2013, vol. 190, no. 11, pp. 1917–1936.

2 I will be referring to philosophy in the following argument, but I see these points as applying across the humanities. The dedisciplining of literature, history, art history, and so on, will take different forms and face different opportunities and obstacles. But each has a vibrant role to play within the public and private sector. I first saw this in the 1990s when I was working with the US Geological Survey. Even more than philosophy, I found that literary studies were the crucial discipline for the Survey, which badly needed to create local narratives for the wealth of information it possesses.

3 The term is used deliberately. The 20th-century tradition of applied philosophy, rather than being an exception to my claims, exemplifies the problems discussed here. See below.

DOI: 10.1057/9781137303028

4 While the situation was quite dynamic, the US Department of the Interior reported in 1890 that in 1886–1887, 62% of college students were still enrolled in classical courses.
5 In an article he wrote for *Newsday* titled "Has Philosophy Lost Contact with People?" W.V.O. Quine notes: "think of organic chemistry; I recognize its importance, but I am not curious about it, nor do I see why the layman should care about much of what concerns me in philosophy." I am indebted to Douglas B. Quine for this citation.
6 Or nearly exclusively disciplinary: cf. Bertrand Russell.
7 A version of this argument was made by Frodeman, Holbrook, and Briggle, 2012.
8 See Chapter 1, note 7.
9 In one of the few philosophical accounts on the PGR that can be found, "Our Naked Emperor: the Philosophical Gourmet Report," Zackery Ernest begins with a quote from Chomsky – and then ironically footnotes it saying: "I apologize to Noam Chomsky for using his quote for such a trivial subject as this one" (Ernst 2009).
10 Elon University is a notable exception, with an innovative program of undergraduate internships for philosophy majors.

Bibliography

Ackerman, TF, 1980. "What Bioethics Should Be," *The Journal of Medicine and Philosophy*, vol. 5: 260–275.

Allenby, Braden and Sarewitz, Daniel, 2011. *The Techno-Human Condition*. Cambridge, MA: MIT Press.

Baron, Jonathan, 2006. *Against Bioethics*. Cambridge, MA: MIT Press.

Beauchamp, Tom L. and Childress, James F., 2008. *Principles of Biomedical Ethics*. Oxford University Press.

Bordogna, Francesca, 2008. *William James at the Boundaries*. Chicago University Press.

Briggle, Adam, 2010. *A Rich Bioethics: Public Policy, Biotechnology, and the Kass Council*. University of Notre Dame Press.

Briggle, Adam and Frodeman, Robert, 2011. "Creating a 21st Century Philosophy," *The Chronicle Review*, December 18, 2011.

Cohen, Avner and Dascal, Marcelo, 1999. *The Institution of Philosophy: A Discipline in Crisis*? Open Court Publishing.

Deleuze, Gilles and Guattari, Felix, 1980. *A Thousand Plateaus: Capitalism and Schizophrenia*. University of Minnesota Press.

DOI: 10.1057/9781137303028

Eckenwiler, L. and F. Cohn, eds. 2007. *The Ethics of Bioethics: Mapping the Moral Landscape*. Baltimore, MD: Johns Hopkins University Press.

Evans, John, 2002. *Playing God? Human Genetic Engineering and the Rationalization of Public Bioethical Debate*. Chicago: University of Chicago Press.

Frodeman, Robert, 2007. "The Role of Humanities Policy in Public Science," in *Public Science in Liberal Democracies*, University of Toronto Press, pp. 111–120.

Frodeman, Robert, 2010. "Experiments in Field Philosophy," *New York Times* op-ed, part of The Stone series, November 23, 2010, at http://opinionator.blogs.nytimes.com/2010/11/23/experiments-in-field-philosophy/.

Frodeman, Robert, 2011. "Interdisciplinary Thinking and Academic Sustainability: Managing Knowledge in an Age of Accountability," *Environmental Conservation*, vol. 38, no. 1, March: 105–112.

Frodeman, Robert, 2013. "Philosophy Dedisciplined," *Synthese*, vol. 190, no. 11, July: 1917–1936.

Frodeman, Robert, Briggle, Adam, and Holbrook, J. Britt, 2012. "Philosophy in a Neoliberal Age," *Social Epistemology*, pp. 18–36.

Frodeman, Robert, Holbrook, J. Britt, and Briggle, Adam, 2012. "Bieberians at the Gates," *Inside Higher Education*, published online on December 10, at http://www.insidehighered.com/

Frodeman, Robert, and Mitcham, Carl (eds), 2004. "Philosophy of Science Policy," *Philosophy Today*, Special Supplement.

Kitcher, Philip, 2011. "Philosophy Inside Out," *Metaphilosophy*, vol. 42, no. 3, April.

Kuklick, Bruce, 2003. *A History of Philosophy in America*, 1720–2000. New York: Oxford University Press.

Lewin, Tamar, 2012. "Harvard and M.I.T. Team Up to Offer Free Online Courses," *New York Times*, May 2, at http://www.nytimes.com/2012/05/03/education/harvard-and-mit-team-up-to-offer-free-online-courses.html.

Light, Andrew and Katz, Eric, 1996. *Environmental Pragmatism*. Routledge.

Longino, Helen, 2002. *The Fate of Knowledge*. Princeton University Press.

Marchand, Philip, 1998. *Marshall McLuhan: The Medium and the Messenger*. Cambridge, MA: MIT Press.

DOI: 10.1057/9781137303028

McCumber, John, 2001. *Time in a Ditch: American Philosophy and the McCarthy Era*. Northwestern University Press.

Moreno, Jonathan, 2005. *Is There an Ethicist in the House? On the Cutting Edge of Bioethics (Medical Ethics)*. Indiana University Press.

Nietzsche, 1886. *Beyond Good and Evil*, Walter Kaufmann, trans, Vintage Books. 1989. Originally published in 1886.

Norton, Bryan G, 1991. *Toward Unity among Environmentalists*. Oxford University Press. Norton, Bryan and Eugene Hargrove. 1994. "Where do we go from here?" in Frederick Ferre and Peter Hartel (eds), *Ethics and Environmental Policy: Theory Meets Practice*. Athens: University of Georgia Press, pp. 235–252.

Reisch, George A., 2005. *How the Cold War Transformed Philosophy of Science*. Cambridge University Press.

Schmidt, Jan, 2010. "Prospects for a Philosophy of Interdisciplinarity," in Frodeman, Klein, and Mitcham (eds), *Oxford Handbook of Interdisciplinarity*. Oxford University Press, pp. 39–41.

Singer, P.W., 2009. *Wired for War: The Robotics Revolution and 21st Century Conflict*. New York: Penguin.

Toulmin, Stephen, 1982. "How Medicine Saved the Life of Ethics," *Perspectives in Biology and Medicine*, vol. 25, no. 4, Summer: 736–750.

DOI: 10.1057/9781137303028

6
Epilogue: An Undisciplined Life

Frodeman, Robert. *Sustainable Knowledge: A Theory of Interdisciplinarity*. Basingstoke: Palgrave Macmillan, 2014.
DOI: 10.1057/9781137303028.

> We commonly do not remember that it is, after all, always the first person that is speaking.
>
> Thoreau

I

The origins of this book go back quite a way. It may be worth while to explain how I came to be a dedisciplined philosopher.

When I was a boy in St Louis in the 1960s my father worked weekends at a bowling alley. The place was called South Twin Lanes. Across my childhood and youth he stayed there from 8 am to 5 pm on Saturdays, and from 9 am to midnight on Sundays. (During the week he traveled as a lock salesman.) Eventually I worked those hours too.

Not that this was work in an oppressive sense. On Saturdays during the school year the lanes were packed with kids bowling in the junior leagues. One of the parents would serve as league coordinator, leaving my father free to operate as overseer and bon vivant. During the summer the lanes could be eerily quiet. Then we'd go out back to play bottle caps – baseball with a broom handle, swinging at the caps from beer bottles collected from the bar. We'd play for hours in the sweltering heat.

Sundays would start out quieter. We would bring in breakfast, often heart-clogging chocolate donuts, which we'd eat in silence while reading the (now defunct) *St. Louis Globe Democrat* – the local Republican-leaning paper, but delivered in the morning and with the better sports page. We'd keep an eye on the few heathen bowlers who would show up during church hours. But basically it was just hanging out until 11am or so.

Then things picked up quickly. During the cold months and rainy weekends the lanes were packed from noon on: teenagers, dates, families. The waiting list would grow to an hour and a half, and would last until 6pm when we would need to clear half the house for league bowling. Then the mixed (couples) league would run until 9 or so. By 10pm the lanes were down to the few stragglers who would keep us there until midnight.

Running the front desk, my father would pass out rental shoes and score sheets – no electronic scorekeeping in those days – call out names from the waiting list, and announce incoming phone calls over the loud speaker (with no cell phones the house phone was the de facto public

DOI: 10.1057/9781137303028

phone). My father would run the cash register, fishing out silver nickels or dimes as they came in, swapping them out with regular coins. He would function something like a cop on the beat, while hanging out with friends, reprimanding the unruly, and teasing young couples on their first dates.

I'd get to do these things too. The situation was loose and entrepreneurial, and I learned how to deal with the public. Most afternoons my father would leave me alone at the desk at some point to sneak into the darkened bar (closed on Sundays, Missouri had blue laws in those days) to catch a nap on the couch. It was in that same bar that I watched Neil Armstrong take his first steps on the Moon while sweeping up the leavings from the previous night's party.

On Sunday evenings after league play I'd put the place back in order. With the lanes nearly empty the sound of strikes would echo through the building. I'd sweep the floor, empty the ashtrays, collect the trash, and run the machine that put down a new slick of oil on each lane. I'd work fast on lanes 1 through 8, and again on 18 to 24, but slacken my pace across the middle lanes. I wanted to hear the stories that weren't meant for my ears. My father would sneak a few beers out of the bar and he and his buddies would talk, or maybe his crew would steal free frames while he cashed out the register. Through it all the stories would continue: the world was a narrative filled with tales of World War II and the Korean War, the Baseball Cardinals, and bowling alley gossip.

At first I collected beer bottles in exchange for free games of bowling. Eventually I graduated to fixing the automatic pinsetters. The machines would mostly run themselves except for the occasional stuck ball or when some clown would hit a sweep. That left me with lots of time to read. I filled the time with Thoreau, Tolstoy, and Dostoevsky; then Eiseley, Stegner, and DeVoto; eventually Woolf, Dillard, and Faulkner.

The reading was a source of amusement to my father's friends. They'd grab a book out of my hands and declaim a passage, exaggerating and mispronouncing the words. They'd point out how goofy and pointless and indecipherable the book was. I usually played along, since it was all good fun, though at other times I'd defend my reading, explaining how the books served as implicit commentaries on our local goings on. They would have none of it, of course. But I grew used to looking for connections between serious thinking and everyday life – between Dostoevsky and South Twin Lanes. It was early training in inter- and dedisciplinarity.

DOI: 10.1057/9781137303028

II

I stopped by the bowling alley recently. The name had changed, but the place looked like it did 40 years ago. Not a good sign. The owner's son let me into the back to take a look at the automatic pinsetters, and volunteered: "The bowling alley business is dying. Young people have better things to do." Bowling on a Wii is evidently now considered more fun than the real thing.

Nonetheless, the bowling alley was preparation for a career. When I went to college – St Louis University, where the Jesuits were well versed in disciplining unruly young men – I wanted to study philosophy. I liked the reading, but I discovered that I most enjoyed discussing philosophy (and history and literature) with science or business majors rather than with philosophy majors. Of course some of the former weren't interested. But many relished the sparring, and made sharp points, all the while offering biting commentary on the worthlessness of philosophy and the humanities generally.

More surprising – although in retrospect it should not have been – was the reaction of my philosophic colleagues. They failed to see the point of talking with the non-disciplinary-trained. Conversations would have to be dumbed down, and concepts would have to be explained. Better to talk to those who were in the know. Being dismissed by non-philosophers made sense to me; but why would humanists, and philosophers particularly, be so uninterested in making arguments about what the humanities could offer to society? Beyond the level of platitudes, that is. Their attitude seemed a betrayal of the first principle – and principal – of philosophy, the Socratic imperative that calls for us to engage people in all walks of life. Socrates, after all, was an inter- and transdisciplinarian *avant la lettre*.

Of course at that point I had little idea about what properly counted as philosophy. I thought of it as the exercise of reason, and of rationality as the raft to carry me through choppy seas. Evidence and logic would link and I would have a net for plumbing the depths. I'd share my catch with others, and they with me. Problems would be solved. But it turned out that the net was rent, and try as I might I could not fully mend it, either by myself or with the help of others. We even disagreed on what was brought up wriggling from the sea.

These difficulties became their own source of interest. Searching for answers led me to read across the disciplines: anthropology, literature,

DOI: 10.1057/9781137303028

history, art history, and geology. And to explore the field of hermeneutics, which seeks to account for differences of interpretation and to understand how reasonable people can disagree. Ironically, this only doubled the problem, as I found that people now disagreed about the nature of disagreement.

While I felt that I was making progress, I was also filled with perplexities that only seemed to grow with further study. This fact suggested something about the nature of knowledge. Endless study, rather than always being a good idea, might lead one down a rabbit hole where uncertainties multiplied rather than lessened. I noted, too, that people embraced an epistemological tic without feeling any need for explanation: they dealt with the confusion of perspectives by picking one discipline (and eventually, a sub-discipline) and sticking with it. This, somehow, was thought of as being 'rigorous'. To me, though, diving into a discipline seemed more like a handy way to avoid the issue of how to reach common understanding across multiple perspectives. The pursuit of depth of learning often led to 'success' at the cost of ignoring the lateral connections – and gaps – between things.

Overall, reasoning and study seemed like fine and beautiful things, but serviceable only as long as you did not push them too far. A central question became how far to push them. The academy seemed committed to taking the whole business as far as time and funding would allow. As an aesthetic this perhaps made some sense, for I too appreciated the pleasures of deep and careful thinking. But as a social practice it seemed odd. Perhaps dangerous. Perhaps even a little mad. The whole knowledge system came to seem out of whack, deep insights floating in a sea of obvious stupidities.

III

Despite these doubts – and in part because of them – I went to graduate school. I studied modern Chinese history, and then switched to Western philosophy. After some trials and tribulations I finished by writing on Hegel and Heidegger, and then snagged a job in South Texas at the University of Texas at Pan American. The weather there was miserable, the culture odd, and the teaching load heavy – four courses a semester. And as the newest hire I was assigned three sections of logic every

DOI: 10.1057/9781137303028

semester. (The fourth was an upper division course of my choice for majors.)

I came in dreading the logic classes. But they soon became my favorite courses: they presented the opportunity to talk philosophy with people who would never take another philosophy class. Tossing Copi's *Logic* aside, we analyzed the logic of commercials, political ads, and film; the logic of the classroom setting, and of the social conditions across the Rio Grande Valley. In the prized upper division courses, however, the vibe felt wrong. Philosophy seemed an escape from the messiness of everyday life rather than a deeper exploration of it – perhaps understandable given the difficulty of life in South Texas.

I lasted two years in Texas. I quit to go back to school in the Earth sciences. I wanted to write an environmental phenomenology of the Grand Canyon and the Colorado Plateau, and felt I needed to know something about rocks. So I enrolled at the University of Colorado. But the master's in the Earth sciences – in paleoclimatology, itself a kind of latter-day natural history – led in a different direction. One day in 1993 the US Geological Survey contacted me. Newt Gingrich and the Republicans were trying to eliminate the USGS, and the Survey was looking for help in demonstrating the social value of public Earth science. Over the next eight years I worked part-time with the Survey to help connect Earth science with social needs – a period across which I learned at least as much as I taught others.

The next ten years flew by in rapid succession: an appointment in Philosophy and Environmental Studies at the University of Tennessee at Chattanooga; a semester at the University of Colorado's Natural Resources Law Center; a year as Hennebach Professor of the Humanities at the Colorado School of Mines; and finally a position at the University of Colorado's Center for Science and Technology Policy Research, where I worked on the ethics and values aspects of policy issues. I had become a jack of all trades, an intellectual utility infielder, and something of an interdisciplinarian. The situation was soft money, but life was good.

Then much to my surprise, for I thought I was done with academic philosophy, I was invited to apply for the position of chair of the philosophy department at the University of North Texas (UNT). Surely this was the only philosophy department that would have considered me for the role. While on the margins of the philosophic community – as defined by people like Leiter – the department's work in environmental ethics

DOI: 10.1057/9781137303028

put it at the center of a number of societal concerns, including climate change. I had found a home.

Four years as chair raised a new set of questions. I was struck by the impotency of good ideas absent an adequate institutional framing. If ideas are to have a public rather than merely private life they needed to be put into practice. And it soon became clear that philosophical practice involved much more than simply 'applying' finished ideas to a situation. This put a new spin on administrative work, a point that I have tried to express through what has turned out to be a somewhat-unpopular coinage of mine, the notion of the philosopher bureaucrat (e.g., Frodeman 2006 and above).

I discovered that institutionalizing ideas required as much thought, and more rhetorical skill, as communication with disciplinary colleagues. At the same time my department's focus on environmental questions meant that we were deeply involved in interdisciplinarity research. Questions of the nature of interdisciplinarity thus steadily grew in importance. What was interdisciplinarity, why was it so difficult, and how did one successfully prosecute it? Why was it simultaneously praised, unstudied, and underfunded?

This led me to the literature on interdisciplinarity. But I was dissatisfied with what I found, despite the many insights offered by those working in the field, for reasons that I have tried to express across these pages. Seeing an opportunity, I approached my provost about creating the nation's first center for the study of interdisciplinarity.

UNT's Center for the Study of Interdisciplinarity was launched in 2008. In the intervening years we have explored questions surrounding interdisciplinarity and transdisciplinarity via a case-studies approach. Across our many projects we have had two overall themes: to develop a philosophy of interdisciplinarity, and to give an account of how philosophy and the humanities can take up a new role in the world today – what we have taken to calling the dedisciplining of philosophy. One result of our work was the *Oxford Handbook of Interdisciplinarity* (2010). Another is the book that you hold in your hands.

As of this writing the future of the center is in doubt. No surprise there: situations and personalities change and economies crater while universities remain committed to disciplinary structures. But similar work will go on in one venue or another. As paleontologists point out, innovations usually come from the margins.

DOI: 10.1057/9781137303028

Bibliography

Frodeman, Robert, 2006. "The Policy Turn in Environmental Philosophy," *Environmental Ethics*, vol. 28, no. 1, Spring.

Frodeman, Robert, Klein, Julie Thompson, and Mitcham, Carl. 2010. *Oxford Handbook of Interdisciplinarity*, Oxford University Press.

DOI: 10.1057/9781137303028

Index

DOI: 10.1057/9781137303028

DOI: 10.1057/9781137303028

DOI: 10.1057/9781137303028

CPSIA information can be obtained at www.ICGtesting.com
Printed in the USA
LVOW13*0835170614

390398LV00007B/98/P